AWAKENING THE EMPOWERED EMPATH THROUGH THE ENNEAGRAM

This Book includes:

The Power of The Enneagram & The Sacred Enneagram

A guide to understand your emotions and develop Empathy and Self-Discovery

ANDY CONNOR

TABLE OF CONTENTS

BOOK 1

The Power of the Enneagram

The Power of the Enneagram

understanding the Enneagram to develop Empathy, Mindfulness and Meditation

ANDY CONNOR

CHAPTER 1

THE ESSENCE OF THE ENNEAGRAM

What is the Enneagram?

The Enneagram is a powerful tool and typology system for personal development based on a belief in there being 9 main types of the human personality. These personality types are known as "Enneatypes," and can further be learned about to explore our own strengths, talents, positive attributes, shadow or negative traits, and inner workings and desires. It is a type of system that describes the human personality in a way which can lead to spiritual awakening and awareness, self- transformation, and self-discovery, whilst simultaneously sparking profound internal shifts and alignment onto our life path or purpose. The idea with the Enneagram is that everyone will be one specific type but can also resonate with any, or all, of the others. In other words, everyone has a specific and unique personality type which further extends into the other personality types. You can therefore have and embody different aspects, traits and positives/ negatives from the other 8 types, yet you will be defined as having one main type. For example, if you resonate with type 9 personality you will be called "type 9."

This unique self- discovery system for personal development, spiritual transformation, and potential collective healing and transformation stems from the Greek "ennea," meaning *nine*, and also Greek "grammos." Grammos literally translates as a 'written symbol,' therefore this can shed some insight into the nature of the Enneagram itself; its is like a self- contained healing system itself, or more specifically- a *nine point system*. The Enneagram is further represented by a *circle* connecting the nine types in a nine-point diagram; and these nine personality types represent patterns of thinking, feeling, perceiving and relating to others, in addition to root struggles, predictions surrounding behavior and emotions, and virtually all aspects of one's life decisions and choices. The significance of the circle should not go overlooked, as a circle often symbolizes the cyclic nature of self, existence and being.

The Enneagram is a modern synthesis of a variety of ancient wisdoms, metaphysical and esoteric schools of thought, and religious and/ or spiritual traditions. Regarding Oscar Ichazo, one of the main figures paramount in the creation and spreading of the Enneagram as a system for healing and self- development, becoming aware of his journey is highly significant. Ichazo was born in Bolivia and raised there and in Peru. He travelled through Asia where he acquired a lot of his wisdom and esoteric perspectives, before returning back to South America to create the foundation of his teachings. This "foundation" was an educational system to all that he learned and after some time it finally became the Arica School. The methods and teachings shared in this Arica School were based on the idea and practice of self-realization, and further includes a range of wisdoms and areas of self- exploration from psychology, metaphysics, cosmology, spirituality and eastern religions. The intention? To bring about transformation and change in human consciousness.

Amongst the teachings of his school and educational framework was the symbol of the Enneagram. The Enneagram symbol itself has its roots in antiquity- classical times and ancient past also birthed from civilization before the Middle Ages. Actually, the symbol can be traced back as far as the times of Pythagoras, and it was reintroduced to the modern world by someone called George Gurdjieff; an inner work school founder who taught primarily sacred dance and movement (with the Enneagram symbol used in each). His pathways and methods of involving the Enneagram symbol were intended to connect participants directly to the essence of this ancient symbol, also allowing them to become in tune with their senses and inner feelings. However, unlike the system of the Enneagram shared today, Gurdjieff did not develop or teach the *personality system* also known as the *Enneatypes.*

***

Enneagram History, Origins & Influences

The exact origins of the Enneagram are somewhat unknown and also a slight mystery, but it is believed that this unique system for self- development stems back to Ancient Greece. This is mainly due to the origins of the word itself. However, it is a synthesis of a number of religious traditions including

Judaism, Christianity, Buddhism, Taoism, Hinduism, Islam, and Greek philosophy. Yet, the Enneagram is not religious, it is more spiritual and esoteric. As briefly mentioned in *What is the Enneagram?*, the roots stem back to classical and ancient times. It can, however, be seen as spiritual due to the innate level of spiritual awakening and transformation which naturally arises with learning about yourself in relation to the personality types. Furthermore, the person responsible for bringing the Enneagram to the Western World, George Gurdjieff, was a mystic (to some extent!). He also arguably acted as a "bridge" or messenger between ancient times and new, and was deeply philosophical and focused on the sacredness of life himself. Although George Gurdjieff is said to have played a fundamental role in the Enneagram's emergence, the finding and creation of the 9 enneatypes was down to *Oscar Ichazo*.

It is interesting to know that Ichazoactually taught a system of 108 Enneagrams in Africa also known as *Enneagons*, however in America only four main Enneagons remain. These are the Holy Ideas, Virtues, Passions and Ego-Fixations, which we explore later. In Interviews with Ichazaro[1] the following was shared by the founder himself:

"We have to distinguish between a man as he is in essence, and as he is in ego or personality. In essence, every person is perfect, fearless, and in a loving unity with the entire cosmos; there is no conflict within the person between head, heart, and stomach or between the person and others. Then something happens: the ego begins to develop, karma accumulates, there is a transition from objectivity to subjectivity; man falls from essence to personality." (Interviews with Ichazo, page 9, retrieved from The Enneagram Institute, 2019.)

Ichazo's perception of the Enneagram being a way to discover and explore the human soul, psyche and self links to its potential and believed origins. The basis of the Enneagram and in particular the four Enneagons described above, tie in closely to mystical and philosophical traditions. Many teachings from Buddhism, Taoism and Christianity share strong similarities with the beliefs and teachings contained in the Enneagram; not the Enneagram as a symbol, but the Enneagram as a system taught today. (The one with the 9 personality types.) The "idea" and concept of 9 divine forms links to Plato's Divine Forms or Platonic Solids, as these are based on the idea of there being essential qualities of existence which cannot be reduced or broken down further. Well, the same is true for the 9 Enneatypes.

Furthermore, Christian mystics throughout time and history have come up with many concepts and advocations regarding the divine and divinity within the ego, or ego consciousness. This can be seen in the "7 Deadly Sins;" similar to the modern- day Enneagrams's "Ego- Fixations" and "Passions." The 7 Deadly Sins of anger, pride, envy, avarice, gluttony, lust and sloth actually began as *nine* forms and variations, yet over time and during their travels from Greece to Egypt reduced down seven. The Kabbalah itself and Kabbalah teachings from mystic Judaism contains the main principle of the ancient "Tree of Life," which is said and known to be a *map* presenting specific patterns into human consciousness and the ego and psyche.

***

The 9 "Enneatypes"

In my other book, *"The Sacred Ennegram: a journey to discover your unique path for spiritual growth and healthy relationships"*, I go through each of the 9 Enneatypes, or personality types, in great detail with a deep exploration into each. We cover how each personality type relates to relationships, career, love, family & friendships, health, spirituality, and virtually all major areas for self- development and life exploration. It is truly a book you don't want to miss. So, in *The Power of the Enneagram,* we will just be looking at the 9 personality types at face value. Please note this section provides only an introduction into the Enneatypes and if you want to explore this subject deeper, you will have to purchase the book.

***

Type 1: The REFORMER

Reformers, or "type 1's," are perfectionists and full of purpose. If you are a type 1, you are self-controlled and strive for integrity in all that you do. You possess a very strong sense of what is the "right" and "wrong" way to do things, and are further strong- principled, idealistic and possessing high values and morals. Other positive qualities and attributes include being honest, responsible and dependable. On a higher emotional level and frequency, you possess great serenity and can let go of things others' may not be able to do so easily.

Type 1 shadow traits? You can be very judgemental, intolerable and uncompromising. Type 1's are also known for a highly critical nature (however this is mainly due to being so idealistic with high morals!) and can fall into resentment and uncontrollable anger if you do not learn how to express yourself. When functioning on a lower emotional frequency, you can become very unadaptable and immoval in your views, which- again- can lead to strong feelings of resentment and criticism.

In saying all of this, if you are a type 1 you highly grounded, practical, responsible and socially- organized and oriented. You can achieve great success when focusing your energies and emotional energy on something with purpose, such as improving the world or societal structures in some way.

***

Type 2: The HELPER

Type 2's are generous, people-pleasing, and relationship- oriented. You- assuming you are a type 2- have a strong desire to be loved, give love, and generally care for others. Helpers/ type 2's value peace and harmony in relationships and often find yourself taking on a mediating or diplomatic role. When functioning at your highest emotional frequency, you possess incredible humility and ability to be humble. Your self- worth and self- esteem is also independent of any faulty perception, judgement or ego delusion.

However, you can sometimes deny your own needs in order to make others happy, being way too self- sacrificing. Other shadow traits include being naive and severely codependent, or falling into pride and slight arrogance. This is usually only a case, however, when you are suffering from low self- esteem and self-confidence.

Despite all of your shadow and negative character traits, you are genuine, empathic, a good listener, and a strong emotional support system for those you love. And these qualities really are second to none. You excel at making connections, empathic communication, and showing real and sincere displays of affection, care and compassion. If you can balance your autonomy and self- sovereignty with your need for connection and, often, praise or approval, you can thrive in any aspect of life.

***

Type 3: The ACHIEVER

Personality type 3's are driven to excel, success- oriented, energetic and pragmatic. You give your attention to fame, success or fortune and aren't shy of appearing in the spotlight, being admired or receiving lots of attention. You are also highly adaptable and generally successful in all you put your mind to. As a type 3, you work well with others and can communicate well. A major strength is in your emotional connection and intelligence, which you use and channel into getting things done. When at your best you exhibit and advanced level of truthfulness and wisdom, and can see beyond the surface and above superficial appearances. Personal authenticity, self- autonomy and self- expression are strong.

Shadow aspects include overly competitive, impatient and not knowing how or when to take a break from work, in addition to focusing excessively on external praise or materialism. Because of your high- flying and high- striving nature, and being known as the "achiever," this need for external praise can sometimes leave you out of touch and tune with your own feelings and which role or persona you should take on. Instead of using your strong sense of autonomy and personal authority to think, act and behave, you allow others to decide what is relevant and important for you.

Despite these temporary "blips," however, if you are a type 3 you are bound for success and personal accomplishment so long as you remember to let go of image and social persona and connect to your inner truth, essence, and emotional wisdom and intelligence.

***

Type 4: The INDIVIDUALIST

Type 4s and individualists and romantics. You can be dramatic, expressive, and forward-thinking. You can also be very emotional, sensitive, withdrawn and too focused on your own feelings, manifesting as becoming self-absorbed and self- centred. You have a strong sense of identity and, like the archetypal "hopeless romantic" and "gifted but suffering artist," you are a sensitive soul who is usually extremely gifted creatively, imaginatively and artistically. But you can also be temperamental and prone to melancholy and moodiness, at times. One of your greatest strengths is the gift of equanimity, which means you have a unique gift for remaining calm and composed even in the most difficult situations. This also implies you have a strong heart chakra or center, and keep your heart harmonious and healthy with loving and positive thoughts.

Due to your extreme romanticism, however, you may often steer towards shadow personality traits; the sense that something is missing from life often leads to a quest for wholeness through the healing arts, mysticism, spirituality or idealism- either romantic idealism or on the humanitarian and charitable wave. You can be extremely selfless yet this selflessness may not always serve you or your highest self. This can further lead to feelings of envy, severe melancholy and closing your heart to others or those closest to you.

Fortunately, a major positive aspect to 4's is in advanced creativity and ability to express universal human emotions through art, music, dance and poetry. You are highly compassionate, empathic, and idealistic with emotional depth and maturity, and this saves you in many situations. Your authenticity, sincerity and empathy are qualities to be valued.

***

Type 5: The INVESTIGATOR

Type 5's are also known as Investigators or *Observers*, and are innovative, intelligent, original thinkers and highly perceptive. You, once again assuming you are a type 5, have a tendency to become detached, isolated, and secretive and suffer from chaotic or difficult emotions. But you simultaneously possess a unique ability to become detached, which allows you to thrive and excel in all matters of logic, reason and problem- solving. Unfortunately, people often misinterpret your quiet and thoughtfulness as arrogance or conceit, and this is what can ironically lead to your isolation or over- intellectualism. Yet, type 4's thrive in scholarly or creative pursuits and can achieve great levels of self- reliance, self- autonomy and personal sovereignty. You focus on accumulating knowledge and thus may become an expert in your chosen field, which can further lead to material wealth and success, positive and harmonious relationships, and success in life.

At a higher emotional level and frequency you are incredibly self- sufficient and self- autonomous. Your advanced and evolved perceptive skills, intuition, intellect, and non- attached nature allows you to thrive in life and overcome any potential shadow, or negative, personality traits. Your challenge and gift when achieved is to participate fully in life and become in tune with your feelings.

***

Type 6: The LOYALIST

Type 6's are responsible, committed and attentive to people's problems. Because you are so loyal, you often enjoy long-lasting and mutually respecting relationships and can be very trustworthy, and devoted, whilst also being aware of what is going on around you. You are very mental and cerebral which leads to enhanced intellect and intuitive thought processes. Furthermore, you have a unique courage which can help you in any situation, and which also makes you appear strong and centered with healthy boundaries and a great sense of self- leadership. The one major strength of type 6- courage- should not go overlooked.

Type 6's shadow aspects? You can be very worry-some! Due to your need for security and connection, you may sometimes become anxious and suspicious or tend to dwell on the negative. You also may become deeply cautious and procrastinate as a result. Despite these traits, as a type 6 you make a wonderful friend, lover and team- player, and often focus on and direct your energy towards community projects and groupwork. You can thrive in a crisis and possess a unique courage and strength which comes with an innate loyalty to self and others. You are attentive, fearless and loyal, and so long as you work to overcome your pessimism and self- created suspicions you can live a balanced, healthy and happy life.

***

Type 7: The ENTHUSIAST

Type 7's, or Enthusiasts, are spontaneous, fun-loving, and versatile. You are forward-thinkers and rely heavily on your inner mental abilities and communication skills. You are extroverted, fun- loving, and social and enjoy staying busy. However, you can also be easily distracted and unfocused. You might start things without finishing them or become scattered and self- absorbed… you have a tendency towards fear of commitment. When imbalanced you can become highly gluttonous, either through food, any type of intoxicant, or through the excessive "consumption" of ideas, beliefs or activities. In other words, "less is more" doesn't hold any significance to you! You can make up for this however through your fun- loving, optimistic and joyful nature. As a type 7 you enjoy travel, adventure and intellectual pursuits and interests.

Type 7's can be deeply inspirational and bring an optimism and unique energy to those around them. You like to keep your options open and are generally not too concerned with other people's opinions or perceptions of you. You are self- autonomous and fun- loving, but this can frequently lead you to becoming self- absorbed, as previously mentioned. Your key in life is to balance and merge your adventure- seeking, free- spirited and quick- witted nature with some grounding and security, or commitment. If you can do this, you can then harmonize your personality with the quality of sobriety, calming your mind and overall aura and learning to become present in each moment.

***

Type 8: The CHALLENGER

Type 8's are bold, dominating, decisive and self- confident. You love to take charge and often thrive in business and entrepreneurial matters. Leadership comes naturally to you and you have a strong inner power, which when channeled wisely and constructively can manifest great change in the world. When functioning at your best, you possess an innocence that allows you to face life with an open heart, authenticity and natural desire to do good. Because you believe in success, personal accomplishment and hard- work- you generally wish idealistic and inspirational ways to be widespread and available to all, and this is where your innocence really shines through. You do not give in to cynics or unhealthy criticism and tend to stay aligned to your own truth and center. You are generous, enthusiastic and powerful.

Their shadow side of your personality is that you can often appear as domineering and aggressive, or at the very least highly confrontational! This "dark" aspect of self may lead you to experiencing frequent outbursts of anger, forced or excessive willpower, and a need for control. In its extreme, this can further manifest as a desire for revenge when someone has genuinely done you wrong, or caused harm to yourself or someone you love. However, every dark has its light, as type 8's will always defend an underdog or use your strength and inner power to protect someone in need. Great success can come to 8's through the need to protect others and live life in an inspirational and powerful way, and combined with your intense energy and drive, ambition and highly-aspirational nature; type's 8's are and can be one of the most successful personality types. In the material and physical world, you can achieve great things and lasting change for humanity.

You are generous, self- sufficient and sincere, but sometimes impatient, competitive and over- bearing. If you can learn to balance assertion and control with cooperation and compromise, you can attune to the better aspects of your personality and self and be seen as the confidence and courageous, compassionate leader that you are.

***

Type 9: The PEACEMAKER

Type 9's are Mediators and Peacemakers, loving harmony, compromise, cooperation and an easy- going life. Defined by an easy-going nature, inner balance and personal sense of harmony, type 9's thrive in any counselling, mediation or diplomacy role. People with this personality type tend to avoid conflict and when using this character trait productively and in an action- oriented way (as opposed to through avoidance or escapism) you can accomplish great things. The type 9 personality is balanced at the top and center of the Enneagram meaning that you have both a highly grounding quality and are often people's gem or rock. You are receptive, agreeable, compromising and understanding, and also highly compassionate and empathetic. In fact, you are literally the peacemaker signifying that you are defined by idealism, a sense of unconditional love, and a desire to do the right thing.

Despite your incredible selflessness and respect for everyone (a result of being so genuinely cooperative, peace- loving and diplomatic) your down- side is that you can become lost in ininertian- a lack of momentum and "moving- forward." You can also give in to complacency, indecisiveness and facing responsibilities, the latter due to an extreme avoidance and dislike of conflict. Other shadow traits include stubbornness, ambivalence and over-sensitivity.

When balanced with an inner harmony within, however, you as a type 9 are incredibly accepting, balanced, wise and perceptive, with an advanced and evolved intuition. If you can learn to transcend your desire to help everyone all of the time and always be the mediator, you can attract life- changing opportunities into your world.

---

1 The Enneagram Institute. (2019). *Traditional Enneagram (History) — The Enneagram Institute.* [online] Available at: https://www.enneagraminstitute.com/the-traditional-enneagram [Accessed 3 Dec. 2019].

CHAPTER 2

BEGINNING YOUR JOURNEY

Intro

The Enneagram has been created and adapted from spiritual and religious traditions, therefore with this comes the Virtues, Passions, Holy Ideas and Ego-Fixations. These are based on the idea of their being a "divine form" and "holy essence" to life itself, and to our core natures. The ***Holy Ideas*** are the higher aspects of yourself relating to the higher mind, or *Higher Self*. The Higher Self is in opposition to the Lower Self, the part of self and personality responsible for our inner animal instincts and passions, and our untamed, raw and primal feelings and emotions. The Higher Self relates to our intuition and connection to the divine. (More on this later.) Each Holy Idea has a corresponding ***Virtue***, which are qualities of the heart symbolized by pure and positive emotions, such as love, empathy, kindness and generosity. When we lose self- awareness and presence, the positive attributes of the Holy Idea transforms into our Ego- Fixation. These are also known as the ***Fixations***. Simultaneously, losing touch with your Virtue can create the associated characteristics of ***Passion***. Passions are our untamed, primal and animal nature often described as the "Lower Self," as previously mentioned, or at least aspects of it. Instinct replaces intuition- a quality associated with the higher self and symbolizing our *gut- feelings* and *inner knowing*- and all aspects of higher connection and awareness, divinity, and spiritual contact are replaced.

It is our innate and fundamental connection to soul that allows us to connect to the Holy Ideas and Virtues, whereas it is our natural tendencies and drive towards our "inner animals," our instinctual, primal and human selves, which equally create the Fixations and Passions. Being aware of these can help you on your journey to spiritual enlightenment, self- development, and personal growth and transformation. Knowing your personality type allows you to increase presence and higher awareness, so you can learn to contemplate and embody the higher qualities of your true self further developing empathy.

Let's look at the specifics in relation to each personality type. The order expressed is HOLY IDEA - VIRTUE - PASSION - FIXATION:

1. Type 1: Perfection, Serenity, Anger... Resentment
2. Type 2: Will/ Freedom, Humility, Pride… Flattery
3. Type 3: Hope, Truthfulness, Deceit... Vanity
4. Type 4: Origin, Equanimity/ Calm, Envy… Melancholy
5. Type 5: Transparency, Non-attachment/ Selflessness, Greed... Stinginess
6. Type 6: Faith, Courage, Fear... Cowardice
7. Type 7: Wisdom/ Plan, Sobriety, Gluttony... Planning
8. Type 8: Truth, Innocence, Lust… Vengeance
9. Type 9: Love, Action, Slothfulness... Indolence/ Laziness

The "Holy Ideas" - Connecting to Your Light

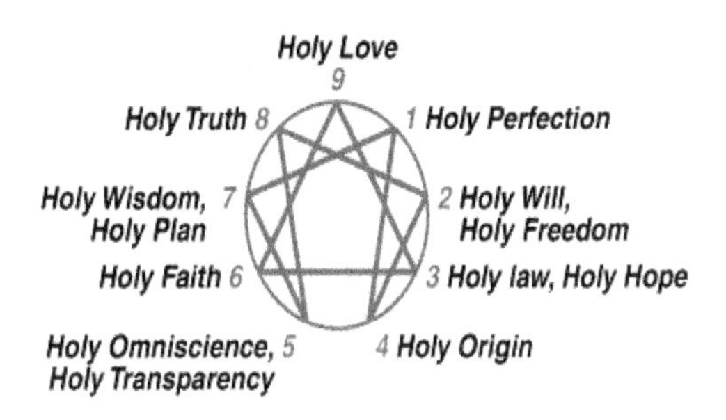

Oscar Ichazo's Enneagram of the Holy Ideas. *Copyrighted Image: "Enneagram Institute."*

Key points to be aware of:-

- Each Holy Idea also has a corresponding Virtue.
- The loss of awareness of the Holy Idea becomes a person's Ego-fixation.
- While we all have the capacity to embody all of the Holy Ideas, only one of them is central to the soul's and our spirit's identity.
- The loss of one's Holy Idea is felt most acutely and profoundly, and

further leads to one's ego becoming preoccupied with trying to recreate it.

- If we lose our center- our Holy Idea- and become distorted in our thinking, feeling, and actions and behaviors, the "lower," "lesser," and shadow/ unhealthy aspects of personality and self start to take center- stage. Thus, we begin to forget our connection to the divine.

"An essential individual will be in contact with these (Virtues) constantly, simply by living in his body. But the subjective individual, the ego, loses touch with these Virtues. Then the personality tries to compensate by developing passions." (*Interviews with Oscar Ichazo, page 19*).

We explore the Holy Idea in addition to the other three, in depth, in How to Use and Connect to the Holy Idea and Virtue to Soften the Unhealthy Side of Your Personality Type.

The "Virtues"

Oscar Ichazo's Enneagram of the Virtues.
Copyrighted Image: "Enneagram Institute."

Key points to be aware of:-

- Each Holy Idea also has a corresponding Virtue, or, in other words, each Virtue will be linked to a main archetype of self known as the Holy Idea.
- The Virtues are essential qualities of the heart experienced by all human beings when we are abiding in "essence," our true selves.
- The loss of contact and connection with the Virtue causes

the person's corresponding Passion. This happens when a person loses awareness and presence, falling away from their true core and becoming disconnected from one's essence.

- While everyone has the capacity to embody all of the Virtues, like with the Holy Idea- only one of them is central to the soul's identity.
- The loss of one's Virtue is felt most acutely, and the person's ego is most preoccupied with recreating it, although in a self-defeating and inharmonious to self way. Again, the Passion is often the result.
- The Virtues describe the expansive, non-dual qualities of "essence" experienced in a direct way, and are further the *natural* expressions of the awakened heart. Virtues are not forced, i.e. we do not force ourselves to be "virtuous." Instead as we relax and become more present and awake, seeing through the fears

and desires of the ego self, the qualities of our Virtue naturally manifest.

The "Passions" - Understanding Your Shadow

Oscar Ichazo's Enneagram of the Passions. *Copyrighted Image: "Enneagram Institute."*

Key points to be aware of:-

- Your Passion arises once you have begun to truly lose touch with your essence, your true self and Virtue.

- The Passion is the primal, "inner animal" aspect of self. To many psychoanalysts and key thinkers in modern times this is equated with the "shadow self," or the key shadow personality traits.
- The Passions are the undesirable, seemingly dark or denied aspects of personality and self. They are seen as in opposition to the divine and represent the lower human characteristic.
- While everyone has the capacity to embody all of the Passions only one of them is central to the soul's and your spirit's identity.
- The Passions represent an underlying *emotional response* to reality created by the loss of contact with our Essential nature, with the core and grounding of our Being, with our true identity as Spirit or essence. The underlying hurt, pain, loss and grief or shame that this loss brings are large and significant, and our ego is compelled to come up with a particular way of

emotionally coping with the loss. This temporarily effective, but ultimately misguided coping strategy which can also lead to many self- destructive behaviors and tendencies, is the Passion.

- Because the Passion is a distortion of an innate essential Virtue, recognizing the Passion can help us to restore the Virtue.

The "Ego- Fixations"

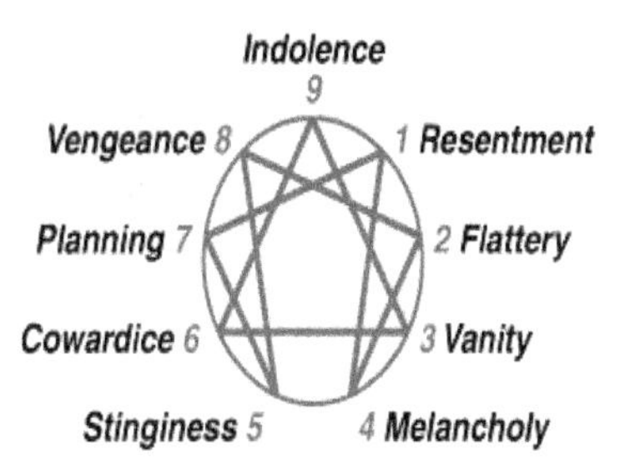

Oscar Ichazo's Enneagram of the Ego-Fixations. Copyrighted Image: "Enneagram Institute."

Key points to be aware of:-

- As the Holy Ideas represent specific non- dual aspects of the core personality (essence), the *loss* of the Holy Idea leads to the Ego- Fixation. The Holy Ideas arise and are embodied when one's mind is clear, calm and in tune with the higher self or mind (or the divine), therefore the loss of/ disconnection from this creates an "ego- delusion" known as this Ego- Fixation.

- Through the Ego- Fixation, one is trying to restore the balance and freedom of the Holy Idea, but due to the dualistic perspective of the ego, cannot. Again, understanding the perspective of our type's Holy Idea functions as an antidote to the Ego- Fixation, so recognizing the Ego- Fixation can help us to restore the Holy Idea.

- As there is a specific relationship between the higher qualities of the self and soul and the corresponding "ego distortions," developing presence and awareness through meditation, mindfulness and empathy can help you overcome the Ego-Fixation and rise to meet your own Holy Idea (and Virtue) within. Your Essence is limited by the Ego- Fixation, just as your Virtue is limited by your Passion.

- Finally (and this is true with all of the subcategories of self) knowing one's "type" is a direct and powerful way to influence inner work to facilitate the transformative process. Healing, spiritual development, and the embodiment of empathy and other qualities in alignment with "essence," can all be expanded through learning about your Personality type.

CHAPTER 3

Empathic Evolution & Spiritual Self-Development

The Benefits of the Enneagram

Let's quickly look at some of the benefits and effects of using the Enneagram for self- exploration, healing, personal transformation and growth before exploring empathy, mindfulness and meditation in depth. Firstly, the Enneagram can help you to become more objectively aware of your personality "blind spots," recognizing your shadow aspects and further learning to overcome them. We all have a shadow, or dark, and a light side- hence why so many traditions, teachings and schools of thought speak of "enlightenment." Enlightenment is literally finding the light within. Yet, just like day flowing into night and darkness and light existing simultaneously, the shadow self is integral to our core personality and inner natures.

Knowledge is power, so the Enneagram brings great wisdom regarding your own flaws and areas for self-development. Since ancient times and times of antiquity philosophers and teachers have spoken of the divine aspect of man, self and psyche, and how transcendental states- or advanced and higher states of consciousness- can be reached. The Enneagram's teachings and aspects of the Holy Ideas and Virtues compliment this, further shining a light onto the inner workings of the soul, self and psyche.

Another major benefit of the Enneagram and learning about your personality type, is that it assists you in developing more self-love by helping you to understand yourself deeper, better, and more profoundly. This can further help you self- evolve, and expand your unique talents, gifts & abilities. Strength and personal empowerment can arise through "*knowing thyself*"- a term and state of awareness often associated with systems for growth and self-evolution. This encourages you to work consciously and with mindful self- talk (self-communication, transparency and honesty) on parts of your personality that may inhibit your ability to live a harmonious and healthy life. Self- love and self- care are the foundations of life, yet without education and self- discovery these cannot be properly applied. Honest and authentic self- analysis, and choosing to see both the dark and light aspects of our personalities equally, accepting them and taking steps to balance them; opens one up to new realms of opportunity and potential for personal growth and self- development.

The personality types, or Enneatypes, are like an omniscient torch lighting up hidden aspects of the self, of yourself. Just like other theories and perspectives regarding consciousness and the subconscious mind, the Enneagram is an important tool for self-discovery. Carl Jung's "Universal Archetypes," for example, are very similar to the ideas and concepts contained within the Enneagram as a Personality identification system. Carl Jung was a psychologist and psychiatrist who believed that we all have shared archetypes inherent within the collective consciousness, and that we also have a subconscious, conscious and unconscious mind. These universal archetypes of the self and psyche can be used to provide insight into our wants, needs, weakness, strengthens, hidden talents and abilities, soulful urges and desires. They can also assist in learning about our shadow- just like the Enneagram deals with the Ego-Fixations and Passions- and thus act as a catalyst for healing and spiritual self-evolution. (We explore Carl Jung's archetypes in relation to the Enneagram later.)

Furthermore, the Enneagram can help you achieve health, vitality and longevity and lead to a unification and harmony of all aspects of self. We are more than a physical body, we have a mind, emotions, spirit and a soul, and we also have subtle versions of ourselves existing. On the spiritual, etheric and astral planes there is a part of ourselves- our true selves- existing and interacting with our daily selves. This is often called and strongly related to the "Higher Self," the part of us which recognizes our connection with the divine and something beyond this 3- dimensional plane of existence. Learning about the Enneagram and your own subsequent personality traits, therefore, opens you up to new ways of perceiving, thinking and feeling. The higher self can be seen as a portal to higher states of awareness, pathways to healing and wholeness, and spiritual awareness and connection to the divine (some higher power, such as God, the universe, source or Spirit). It also connects you to your psychic and intuitive abilities and senses, and is subsequently a link to all spiritual or "supernatural" abilities. Precognition, the "4 clairs;" clairvoyance, clairsentience, claircognizance and clauriadience; intuition, telepathy, the ability to perceive subtle energy, and dream states. When one is in tune with the higher self, they are also attuned to their seat of inner guidance, "gut feeling" and internal compass. (Otherwise known as intuition!)

Linking to the last point, the Enneagram can also help align you with your true path, soul service, or sense of life purpose and destiny. We all have a calling in the world- a unique "soulprint"- a perfect vibratory blueprint of our essence and self. It is important to note that "destiny" here is not seen as something completely above and beyond us, or the human self. The human experience is a combination of both light and dark, shine and shadow, and the higher self and lower self. Your "soulprint" or unique blueprint of self is these aspects balanced, integrated and harmonized, and this is what the Enneagram aims to achieve. Once this has been attained and the dualistic, yet complementary forces embodied, this is where you can truly begin your journey integrating your Personality type; and the subsequent purpose or life path involved with mastering yourself. Overcoming our shadow aspects and integrating our strengths and qualities is an essential step to attaining this. New heights can be reached, new opportunities can arise, and new and advanced or evolved levels of awareness can be attained. As the Enneagram teaches, we have a primal, inner animal and ego- driven self. "Destiny" is the holistic human experience in its entirety, and understanding the quality of *empathy* and how it can be enhanced, developed and embodied- through the Enneagram's teachings- is one sure way to aligning you onto your life purpose and true path.

Finally, through learning about your core struggles, passions, holy ideals and virtues you can come to terms with those parts of yourself you may not want to accept, and the most brilliant and beautiful aspects you have yet to fully embody and balance. Spiritual self-development and self- mastery, which is otherwise known as mastery of your mind, body, emotions and spirit, can only be experienced once we have done the "inner work," got completely honest and real with ourselves, and treated our own lives with authenticity, open vulnerability and an utter willingness to learn and evolve. Of course, we also need to be open mentally and on a soul or spiritual level to the possibilities and dimensions of spirit, or spiritual awareness and knowing. The fact that having or resonating with a "type" doesn't mean you are boxed into one personality, ie. you can be one type but share aspects of another, or many others, simultaneously; means that resonating with a type signifies the tendencies from that personality manifest stronger, and that you are more prone to the positive and negative aspects than the other Enneatypes. So, this in itself gives you key clues and prompts regarding significant life decisions, correct routes and pathways to choose- most in alignment with your soul and spirit- and where you can naturally thrive and shine. (And simultaneously where you may potentially fall or fail.) Your type is you "basic personality type" acting as a foundation and core aspect of your true nature.

***

How to Use and Connect to the *Holy Idea* and *Virtue* to Soften the Unhealthy Side of Your Personality Type

In this section we will look exclusively at the *Holy Ideas* and *Virtues* of each Personality type, and how they can further be used (and connected to) to "soften" the unhealthy side of your personality. This, of course, relates to the *Passions* and *Ego- Fixations* which can be seen as unhealthy or "shadow aspects" or your own Enneatype.

Type 1: Holy Idea: PERFECTION… Virtue: Serenity. Overcoming/ softening, Passion: Anger… Fixation: Resentment.

If you are a type 1 you are a perfectionist with a strong sense of idealism. You want to be right and act and behave in a moral way, and your strive for higher morals and ideals in all that you do. You also possess incredible integrity- more so than most; this allows you to stay true to yourself and remain self-empowered, confident and centered when other people's ignorance, judgments or delusions try to interfere with your centeredness, or "put out your light." At this stage, it is important to remember that a loss of essence with your Holy Ideal, so in your specific case- your perfectionism and idealism, can lead to an increase in your ego- Fixation; resentment. Simultaneously, a loss of connection to your Virtue serenity leads to an increase in in your Passion, anger. Thus, aiming to increase and develop your Holy

Idea and Virtue, and your other positive personality qualities, to the point of integration and perpetual embodiment, allows for the natural softening and decreasing of your shadow personality traits.

The key strengths and personality traits we will be looking at are:-

- Honesty
- Responsibility
- Morality
- Integrity
- Idealism

Honesty

Due to your Holy Idea of perfectionism, signifying that you need to attain the highest and strongest sense of self- mastery and perfectionism, within and in all you do; you are extremely honest. To you, even a white lie can take away from your moral integrity and personal ethics, and this trait in itself can lead to incredible levels of trust, connection and platonic- intimacy in friendships and close relationships. It can also assist in softening your tendency towards anger (short- term) and resentment (long- term) which arise when you are out of balance. Inner balance is essential in life, and harmony comes through an integration of light and dark- the light within and your shadow traits and personality characteristics. We *all* have a shadow self, and this is something which can be hard for you to accept, especially considering your Holy Idea. (Perfectionism!) So, through seeking to increase honesty, one of your strengths, you start to accept the parts of yourself you may wish to deny, or reject. Honesty is not just about being honest with others; it is also about being honest with yourself and this extends to accepting your flaws and follies.

Honesty can be cultivated through mindful communication, empathy, spiritual or self-healing practices, such as meditation, and a variety of therapy modalities. Sound therapy can significantly improve your mental and emotional well- being and health, which further increases your capacity and willingness to be honest. Serenity is your personality's Virtue, meaning that despite how hard you may be on yourself, or others, sometimes due to your incredibly high sense of idealism and perfectionism, you generally have a serene and calm demeanour and outlook on life. This is one of your greatest assets! By being conscious of your inner serenity, appreciating it, and seeking to embody more of it in daily life, you will start to notice how your Passion and Ego- Fixation naturally diminish. Anger builds up when you lose touch or awareness with serenity and accompanying characteristics; calmness, composure, peace of mind, tranquility… so, why not create a conscious reminder to incorporate self-development exercises surrounding your virtues, into daily life?

Responsibility

Responsibility defines you. You are practically- minded, dutiful, in control of your life and accountable- and you expect to see these traits in others too! In fact, when you come across people who lack all grounding and responsibility you can become very frustrated indeed. It may even lead to real displays and outbursts of anger, and over time this leads up to resentment. (Are you starting to notice the innate pattern between a loss of awareness with the Holy Idea and Virtue, and Passion and Ego- Fixation? Good.) Because of your highly grounded and dutiful nature, you often create much pressure within yourself to be right or perfect, or possess the most supreme morals and ethics. This is great, for the most part, but this pressure can also lead to a lot of physical pressure manifesting as health problems. Stress and tension literally store in your body, such as through back, neck and shoulder pains and muscle tension. Over time, this stored tension adds to your ever- growing Ego- Fixation- your natural tendency towards resentment- and further increases your stored and often repressed anger.

To expand on this last point, one of your major defense mechanisms is to avoid and repress anger and other associated feelings, is the desire to appear "right." Quite simply, anger is part of your problem- or at least your perception of it. You see anger as wrong, bad and even "evil"- once more due to your perfectionist motivations- and therefore the very idea of showing and expressing your angry feelings and sensations makes you question your morals, heart and integrity. It is as if you create a war- path with yourself and this ultimately keeps the destructive cycle continuing round and round. Holding such a strong belief that feeling angry is wrong, or makes you less of a person, is what leads to the increase in stress and tension and this further perpetuates the hidden and increasing anger. Fortunately, recognizing this cycle and seeking to "nip it in the bud" early on can lead to great growth, personal healing and transformation, and evolving past your need to turn towards egotistical tendencies (resentment).

Two other profound ways to soften and overcome the shadow sides of your personality is to, 1- avoid *over- committing*; and, 2- be less critical of yours and others' mistakes, downfalls and imperfections. Placing so much emphasis on responsibility, duty and obligations can create this "constant strive" inside, a strive which may lead you to over committing and taking on too much. Of course, this leads to angry feelings; anger at yourself for the stress and pressure you are causing yourself, but also anger at those who don't hold the same values in responsibility as you do, and who further may literally be not pulling their weight. Linked to this is the fact that your perfectionism creates a negative vibration with you sometimes- you may actively show your anger and frustration, or resentment, to others if they are not living up to your extremely high expectations. This can cause many problems including being or feeling alienated, being unadaptable and so rigid in your opinions that you push others away, and diminishing your self- growth. If you can learn to be more accepting, understanding and forgiving this will help you steer towards your virtues and strengths, and leave behind your shadow tendencies for good.

Morality

So, you are deeply moralistic and ethical with high standards; standards of self, standards of others, and standards of how the world and society should be. You may choose a career or path dedicated to justice, morality and helping others through wisdom, truth or knowledge of self; or you may equally become an activist, charity or humanitarian worker, healer, or legal person. You are full of purpose and principles and because of your self- controlled and serene nature (when not falling into your shadow or egotistical tendencies), you have considerable capacity to inspire or educate others. If you do not choose a career in alignment with your morality, you can still be a wayshower and inspiration in your family or small circle of friends and peers. Now, in terms of softening your shadow (unhealthy/ counterproductive) personality traits, having sincere morals and principles in life can be your saving grace. Even if you lose your cool and calm and let your inner anger rise, you still more times than not get your point across and appear in a positive- and attractive and inspiring- light. This is because you are full of *passion*, passion for justice, an ethical way of living, and respect and care for others. A lot of your fire is fuelled by both your integrity and your need for justice, regardless of how small or insignificant something may appear.

You have a gift for seeing the big picture, and you are incredibly wise, perceptive and noble in your energy and intentions; and this makes you a change- maker. Any possible glimpse of what you believe to be evil, wrongdoing or injustice can have your inner warrior burning bright. This does lead to anger of course, but depending on how you express this can help you overcome feeling resentment. Remember, the universe exists in a state of duality and opposing yet complementary forces. Light and dark, yang and yin, white or light and black… these are inherent within ourselves, and within life itself. So, becoming conscious of your inner passion and fire (literally, your Passion) and learning how to express this in a way which heals, creates and constructs, allows you to shine and speak your truth whilst not losing touch with your strong sense of morality. One of your key motivations in life is to strive for something better or higher, and to improve the world around you or the lives of others. Becoming mindful, and perhaps merging *mindful empathic communication* with your speaking style, will open up new doorways to connection, compassion, and constructive avenues to be the change- maker you know yourself to be inside.

Integrity

Your integrity can be awe- inspiring, truly. You possess the advanced and unique qualities of being wise, discerning, realistic and noble. You also have such a high level of integrity that others will see you as a wayshower and inspiring, even when you aren't aware of how they are feeling. This is because you are also humble- you are not arrogant or egotistical in your integrity, wisdom and awesome qualities. Unlike some other personality types, your egotistical or unhealthy ways aren't falling into pride, deceit or vanity; you do what you do and say what you say because you mean it, and they are your true motivations and genuine feelings. Your downfalls, however, lie in your tendencies towards anger and resentment. So what does this say about you and your character? It says that you are highly passionate and full of integrity, honor and positive righteousness. Honesty is highly valued to you and you seek wholeness, unification and cohesion in all you do, but this also means that these favorable traits and qualities can only be shown and expressed when you are conscious of your "Passion" and "Eg- Fixation."

It is no use having such beautiful and brilliant qualities if you choose to let the unhealthy sides of your personality get the better of you. It is essential, therefore, for you as a type 1 to seek harmony, inner serenity and self-acceptance, and these can be achieved through a regular self- development activity or exercise and self- healing.

Idealism

You are one of the, if not the most idealistic personality types of the Enneagram, which means that you have a sense of mission, purpose and higher values. You may very well be in tune with your higher self, or connect to the divine daily, such us through developing and integrating the qualities of intuition, spiritual awareness and recognition, compassionate action, and empathic service. Speaking of service, you are deeply service-oriented and this can make you a gem for the environment, animals, specific groups of people, or anyone on the receiving end of your need for justice and change. Your challenge is to accept what cannot be changed, and develop the wisdom and discernment to know the difference; i.e. are your efforts and energies really being put to the best use? If the answer is no, this may lead to some of your angry and resentful feelings and tendencies, and who does this serve? The answer is, clearly, no- one- it doesn't serve you and it certainly doesn't serve the people you are trying to help, assist or inspire. One of the most effective and beneficial things you can do, for both your peace of mind and to increase your Virtue of serenity, and for staying true to yourself and keeping you admirable strengths and positive personality characteristics strong and in play; is to *learn to accept and self- forgive.* Being less critical of others may sound so simple, yet for you it is a massive step towards healing, wholeness and integration. This is very important, so do take note.

Finally, asking others to accept some responsibility and actively *delegating* will help you overcome any resenting feelings, and help control you ever- increasing injustice fused anger in the present. If someone does not wish to learn or respond to your rightful and equality- centered requests? Simply walk away. Not everyone wants to change or will change, and accepting this fact alone is what is required for your peace of mind and sense of serenity. It further helps you be your most brilliant, idealistic, integral and perfectionist self, as you can continue to be you with your morals and values, wisdom and perceptiveness, also without "forcing" others to change. (Or getting angry at them if they don't.)

To conclude, one thing you need to be mindful of is the need to justify yourself, even if sometimes. Some people just don't get you, and this is fine. A key teaching for yourself is to learn that people with always have their judgments, illusions and certain ignorances- and this is OK; we are all human, after all. If you can accept this you can start to truly embrace and integrate your strong sense of perfectionism and idealism, this in turn will help you soften your shadow, unhealthy and unfavorable characteristics. To do this, therefore, it is important for your true nature to combine perfectionism (your Holy Idea) with *serenity* (your Virtue). This will help you achieve states of inner calm and peace which further helps you to overcome and soften your Passion, *anger*, and *resentment*- your Ego-Fixation. Serenity is your personal key to self-growth, enlightenment, and leading a healthy, balanced and unified life. Furthermore, learning to accept that your strong sense of idealism, integrity, morality and honesty, which link closely to your perfectionist personality, are not right nor valued by everyone, enables you to overcome those angry and resentful feelings you may sometimes experience. In essence, your mental, emotional, physical and spiritual health are extremely important; so honor this.

Type 2: Holy Idea: WILL/ FREEDOM… Virtue: Humility. Overcoming/ softening, Passion: Pride… Fixation: Flattery.

Your key strengths include:-

- Generosity
- Service
- Empathetic
- Sincerity
- Warmth/ friendliness/ caring

As a 2 type personality you are a helper at heart. You value honesty, sincerity, friendship and close connections, and your are also deeply empathetic. Your compassion can be inspiring and you possess a unique and rather extraordinary sense of unconditional- love. Nothing is too much for you and, when at your best, you are humble and genuinely happy to serve in some way. However, with any light there is dark, and you have some things you need to be mindful of. Becoming conscious of your need for wanting to be loved, appreciated or noticed can help you avoid being overcome with your Passion; *pride*. Equally, learning to not need *flattery*- your egotistical tendency also known as your Ego- Fixation- will enable you to speak and act from a place of sincerity, humility and real empathy, which are your greatest strengths.

Generosity

Your Holy Idea is freedom, or will, and with this is accompanied your highly generous nature. Known as "the helper," if you possess this personality type you have a massive heart and nothing is too much for you. Those who love you and you love equally are more often than not on the receiving end of your kindness, giving and acts of selflessness; there is nothing selfish about you. In personal and family relationships you shower those closest to you with love, affection, support, gifts, warmth and friendship, and are a rock or gem in their lives. Your natural tendencies towards giving and generosity are accompanied with your main virtue- humility. In other words, you don't seek praise for your gifts and have a strong sense of self- worth. However, as with every personality type we all have a shadow self and a tendency to steer towards unhealthy habits. Yours are pride and flattery.

Ultimately, learning to do things from a place of sincerity will enable you to overcome all feelings of needing or wanting external praise, recognition and admiration. It will also allow you to develop self- esteem and your self-worth even further so you will never need to be liked; you will do things simply because of your warm and beautiful heart. Losing touch with your Holy Idea signifies that you no longer feel free to be yourself. This may be due to weak boundaries, always feeling like you need to give, or as if should be going out of your way to help others constantly. You also may tend to feel trapped when there is little balance and inner harmony in your life, and subsequently turn towards your Ego-Fixation; flattery. Thus, the once selfless, wholy generous and humble version of you transforms into someone who craves constant attention, praise and approval for their seemingly "selfless" acts. This loss of awareness with your inner strength of will, which can also be seen as your *intentions*, and feeling free to be yourself leads to a reality in which egotistical and unhealthy tendencies take over. It is almost like you "act out," and instead of being wise, empathic and deeply mature and helpful in your emotional dealings, you instead become immersed in this child- like energy and frequency.

The best way to deal with this and "soften" this self- destructive part of your personality is to engage in healthy boundary setting. Being centered, self- aligned and empowered within allows you to be true and sincere in all of your dealings, whether these be emotional, physical/ material or spiritual. The more you can overcome your tendency towards unhealthy pride, the better able you are to connect to your deep and brilliant quality of humility, and this then changes the cycle so you can remain centered and aligned to your true self. (Your *Essence*.)

Service

You are one of the most service- oriented personality types in the Enneagram and this makes you incredibly likeable and loveable, naturally. People appreciate you and generally see your intentions and motivations as pure. At your best you are wholly unselfish and altruistic, however when you lose touch with these qualities you can become extremely self-sacrificing. This means you sacrifice yourself for others or some cause. This may be your mental health, emotional well- being, physical resources, time, energy, love, spiritual harmony, wisdom or personal power; however it manifests and in whatever way, you sacrifice it. This then leads to the embodiment or your unfavorable characteristics, pride and flattery. The more you engage in unhealthy self- sacrifice, the more you fall into personal traps and cycles; and not only does this perpetuate the cycle of needing to be liked or praised, but it also can lead to mild resentment and anger. This form of "disintegration" can create a version of yourself very much like the shadow aspects of type 8- you may become aggressive, domineering and over- bearing, with a bit of a "bullyish" nature. Fortunately you can recognize this tendency with the help of the Enneagram and take steps towards positive integration.

The results? Self- nurturing, empathetic, sincere and selfless; *in moderation and balance.*

This is the key here for your personality type. Because you are so empathic, you truly do forget to take care of your own needs at times, and not only does this lead to the problems explored below in the next area of self-discovery but it can also lead to being naive and harmfully codependent. Service should not be confused with slavery or servitude, therefore looking towards you Virtue humility and your Holy Idea will (will power, freedom to be yourself, and sincere intentions/ motivations) will open you to a reality where you feel self- empowered, expressive and genuinely empathic, and can remain true to yourself whilst not falling into entrapping tendencies (seeking constant approval, pride, etc.).

Empathetic

Emotionally aware, evolved and connected, as a type 2 you are deeply caring, compassionate and giving in your emotional support and wisdom. You are equally highly emotionally intelligent, meaning that you operate at a much higher emotional frequency than most people. You excel at making connections and whether in friendship or business relationships, you are someone many subconsciously and consciously look to to keep bonds in tact and thriving. You act like "emotional glue" and this makes you respected and cherished, even if people don't always express it. There is an "unspoken bond and silence" with you, your vibration and your demeanour, and people know they can turn to you in times of need. You are an incredible listener and give sage advice. And this all stems from your evolved empathy. Empathy is the ability to know what someone else is feeling, as if you are literally in their shoes. But empathy extends beyond this- it also symbolizes the ability to know what others are thinking, sense hidden beliefs and impressions, and sense what state of health someone is in. This can be physical health in that you actually know when something is wrong, such as through an ailment or illness, without being told; or mental, emotional or spiritual health and well- being. The best way to picture empathy is like dolphins who communicate telepathically through supersonic radar, or snakes who have an air of mystery and sense vibrations through their tongues.

So, in terms of your true self and personality, you are deeply empathic and a major support system. However, with any shine there is shadow and this signifies that you have a tendency to act as an *emotional sponge.* Essentially you pick up on everyone's "stuff," their thoughts, feelings, worries, concerns, fears, problems, relationship or career baggage, illness and inner currents- and this can leave you completely drained. If you fall into a cycle of developing and integrating the negative effects of empathy, you can start to really suffer and this in itself can lead to unhealthy habits and behaviors. You need to be careful of absorbing everyone else's problems because, if you don't, you may be prone to emotional outbursts. These emotional outbursts can leave you in a very negative and ugly light, further taking away your essence and all the beautiful qualities of your personality. You may have spent years growing and developing a friendship, where you are a shining and guiding light full of love, compassion and higher helpful perspectives; but in just one or two moments, it can all come undone. There are two simple yet highly effective ways to overcome this. These are, 2- setting personal boundaries, and 2- paying attention to your own feelings, needs and inner currents.

Once you begin to do this, you will begin to notice a massive internal shift which allows you to be humble, sincere and helpful without the need for praise or approval. Your empathy will come from a place of empowerment, self-worth and wisdom, and people will respond positively, which in turn increases your capacity to feel and exhibit so much empathy, and to feel free and self- empowered (your Holy Idea).

Sincerity

When at your prime, you are sincere. People value your time, energy, wise perspectives, heart and soul. In fact, you are the only personality type to display such a strong sense of unconditional love. You have a warm glow in your heart and people are drawn to you, you inspire them and make them feel self-empowered or self- worthy, for example you may often bring out qualities inside which they forgotten existed, or never even owned and embraced. Your altruistic and giving nature is an asset in many people's lives. Yet, your inner self- development can become very limited by your Passion, pride. Connected to pride is your shadow perosnality's traits of self- deception and becoming overly- involved in others' lives. Quite simply, you can be deeply meddling and over- bearing. This, of course, does arise through a genuine, humble and generously selfless nature, however it can still leave you in some tricky spots all the same. People may start to question your intentions, or they may feel trapped, cornered, inferior or suppressed. When you lose tune with your essence, your will and your humility, manipulations can begin. This may be emotional manipulation through your naturally advanced and developed empathy, but it could also include any type of manipulation such as material/ physical or in the form of "mind- games." If you are not manipulating others, you may be entwined in some major self- deception.

Transformation and spiritual self- evolution involves recognizing and accepting the dark parts of yourself so they can be healed and transcended. Due to your lack of boundaries and advanced empathy, this can be very hard for you- and more so than the other personality types. Your empathy means that you are always picking up on other people's emotions, problems, shadow traits and concerns, and seeking to help them through acceptance, love and compassion. Your lack of boundaries means that you sometimes have difficulty knowing where one "self" ends and the other begins, so, in short, when it is time to look to yourself for healing and self-development; everything may be a bit of a "blur." Your ability to merge with others and unconditional love on deep level ultimately implies that seeing our own shadow and "dark" o unhealthy aspects may be shielded. Why and how? Because it's all love! (Well, that's the heat- warming and compassionate advice you give others anyway, and it truly is empowering and helpful.) But, when it comes to yourself this is not healthy for your self development, well- being, or emotional connections with others, and over time this can further lead to a feeling of worthlessness, low self- esteem, and issues in the service-selflessness/ self- sacrifice balance.

Through learning to be yourself and treat yourself with the same kindness, care, love and compassion you show others, and by being mindful and conscious of falling into flattery or egotistical pride; you can achieve the innate levels of wholeness, balance, harmony and love you secretly desperately desire.

Warmth/ friendliness/ caring

Soul, heart, passion, empathy and spirit define you. You are not like most people as your selflessness and service can be truly awe-inspiring; however, you are not free from your follies when it comes to your positive strengths of being warm, friendly and caring by nature. You have this hidden feeling and desire that you must do something extraordinary to "win over" the people you love, or those you want to love you. Although you are not insincere by design, you still possess the capacity to continuously strive for something above and beyond your human capabilities, and as if you must be an actual superwoman or man. In short, you have enormous expectations and emotional needs which can often be repressed. These repressions, unfortunately, lead to the shadow tendencies of losing yourself, lacking boundaries and seeking praise or validation. Furthermore, the immense pressure you put on yourself (often unknown to the people who do appreciate and value you) leaves you feeling insecure, super- sensitive and vulnerable; which further exacerbate your Passion and Ego- Fixation. There is a reason why the Passion is called such; it is your passion, your inner fuel, fire and desire which, as established, is wonderful when expressed healthily.

Although your warmth, friendliness and care is mostly authentic your hidden repressions of feeling like you always need to put others first, in all that you do, leads to many self- harming and unhealthy side- effects. Tensions can build up manifesting as physical symptoms, or muscular aches and pains; inner resentment can build, inevitably leading to explosive emotions as stated earlier; suffering ill- health can arise from always being “everyone’s empath;” and issues in codependency, manipulation tactics, insincerity/ inauthenticity, being domineering or bullyish, or playing victim… these are all common in a type 2 personality when losing touch with essence. Finally, in your desire to hold the self- image of being “helpful” (you are known as the *helper*) your perception or yourself can become faulty, distorted or self- deluded and this in turn can create issues in your close relationships. Luckily your strengths can be maintained with *conscious awareness* of your negative character attributes, *self- love*, *self-compassion*, and seeking to embody and integrate *balance* in all you do.

Your challenge and key to perfect health is simple: find oneself in relationships, balance dependency and autonomy, and practice self-love.

Type 3: Holy Idea: HOPE… Virtue: Truthfulness. Overcoming/ softening, Passion: Deceit... Fixation: Vanity.

The strengths for Type 3 are as follows:-

- Charming
- Successful
- High- Achieving
- Driven/ ambitious
- Inspirational (A "Role Model")

Charming

You are incredibly charming, sociable and liked and these are good qualities to possess. For the most part, or when at your best, this charm is natural, sincere and authentic- there is nothing dishonest about you. And this can be seen and reflected in your Virtue; *truthfulness*. Because you are highly passionate in life, passionate about success, public or self- image, the way you are perceived by others and ambition, you can sometimes be prone to conceitedness. In other words, you can become extremely vain, self- centered and sometimes narcissistic. This, however, only happens when you lose touch with your true self and with your essence. Your fears of failure or "public humiliation" can bring out the shadow and unhealthy side of your personality, which is the need for admiration. Narcissistic tendencies, unhealthy self- love, excessive pride and deception cna result, and if or when they do the persona you seek so hard to create and maintain can become damaged. Fortunately, being aware of this can lead you to seek to embody a natural and graceful charm and be charming through *positive* reinforcement. (And a positive self/ public image!)

Taking steps towards healing and integrating your favorable quality of charm is highly important, and this is reconfirmed when looking at your unhealthy, and often dark, character traits. Because of your high- flying nature, any perceived or self- believed aspect of failure or wrong- doing, on your behalf, can be masked by terrible deception and falsities. Unlike some other personality types who aren't ashamed to admit when they have been wrong, you have a strong and almost instinctive aversion to this. Any thought of your imperfections and flaws being exposed can leave you turning towards some pretty devious behaviors and actions. You may start to become truly untrustworthy in your intentions, actions and interactions with others, or you may become malicious and have no problem betraying those closest to you, or with your best interests at heart. Delusions and jealousies can also arise from your innate fear and strong disliking of appearing anything but brilliant and "above the rest." Finally, your once beautiful and cherished charm may transform into an ugly arrogance, again displaying itself as a form of narcissism.

To overcome these tendencies and "soften" these aspects, meditating on the meaning of charming can really help you. We often lose touch with who we truly are, at our cores, when we become so wrapped up in some external or superficial reality. Because your strengths lie in the physical realm, in a material and often public reality, this is more so true for you. So, to soften this and put you back in touch with positive and amazing aspects of your personality, engage in frequent visualization and mindful meditation. Bringing past memories and visuals of when you were at your most charming will help you deeply, and remind you of when you were at your most successful and connected to the world around- and meditation can really aid with this. Incorporating a daily "charming mindful meditation" will enable you to jolt your memory each day, keep you aligned and on track, and center you with a success and driven- focused ambition; combined with poise, grace and humility. There is nothing wrong with being charming and inspiring those around you, but doing so in the *right* way- without falling into narcissistic or manipulative tendencies- is both your key and challenge in life.

Successful

We have already established that you are defined by success. You wish to impress others with your superiority, knowledge, achievements and talents; you enjoy promoting yourself and putting yourself in the limelight; and you are self- assured and energetic, with a high and healthy self- esteem. Quite simply, you need to be and feel successful in life and you won't settle for a lifetime of selfless service or extreme humility. In saying this, you are deeply charitable, modest and sincere in nature and personality, and you certainly don't lack heart. In fact- you enjoy inspiring others and using your own success to help others in some way. This is an admirable trait to possess; yet, like with all Enneatypes you can become imbalanced and out of tune with your Holy Idea, Virtue and other encompassing virtues. When you begin to lose awareness of self and of your higher self in particular, you begin to fall into egotistical traps. Your hope becomes diminished falling into vanity, and your truthfulness and authenticity diminish into deceit and betrayal; betrayal of your path, of your peers or followers, of your friends and family, of your service, and of your true self. You can regain your hope and inner vision, your Holy Idea and ultimate state, by seeking to overcome your shadow tendencies.

And this can be done in a number of ways. Firstly, you need to keep your intentions and motivations pure and on point, and this can mainly be done by working towards a better world. You will aim to be successful in all you do regardless, so why not be successful in helping a group of people, humanity as a whole, the planet or an animal species? This is a very important piece of guidance and insight to keep in mind, and especially considering you love being in the limelight and appearing in a favorable light. Any act or service which helps you to be successful and be seen in a positive and beautiful light *naturally,* will help you soften your vain and deceptive side. How can you be full of deceit and conceit if you motivations for success are pure? Well, the answer is that they can't. This in itself can also help you with one of your other unhealthy attributes, and that is of becoming *image-conscious.*

You take initiative and work hard to accomplish your goals. Not only are you are a hard- worker, but you also channel a lot of emotional energy into your personal success; which essentially means that you are not so materialistic that you don't value, need or appreciate authentic connections and bonds. In fact, a lot of your motivations and ambition in life stems from your need for emotional connection and significant relationships. Shifting your soul self, service, career choices and life path into something helpful, positive and inspiring, therefore, allows you to overcome and heal any potential you have for becoming overly image- conscious. If you are feeling genuinely connected to others, and further attracting people into your orbit on your wavelength, there is no need to question your intentions or your success. Success is a state of mind, heart, spirit *and* soul; it is not just a physical thing. In terms of your inner hope and truthfulness (yout Holy Idea and Virtue), balancing and integrating these into the core of your personality- through your life choices and success- oriented motivations- enables you to more easily and effectively transcend the need for unhealthy praise and approval, and the turning towards deceitful, manipulative or phony behaviors.

High- Achieving

At your best you are pragmatic, inspiring, relatable and impressionable. You believe in the power of community, society, authentic connections and culture. You may also be highly interested in charitable or humanitarian causes and concerns. In short, your high- achieving nature is by no means inherently superficial, overly- materialistic, or detrimental to your heart and inner essence. You are also self- accepting, gentle and benevolent when connected to your true self and Essence (divine/ ultimate nature), and this expands your ability to be successful in life. Accomplishments and achievements may come to you in abundance and furthermore you often have the support of many, or at least a few close friends, family and partnerships, who truly mean something to you. And this is where you inner hope comes from. You truly believe in the prospect of a better world, and combined with your heart and benevolence; the world is your oyster.

Taking time to *energize* all of these fabulous qualities on a daily or regular basis can be your saving grace. You have strong natural tendencies to become vain, superficial, shallow, deceptive, narcissistic, manipulative, competitive, and even vindictive. Thus, energizing your strengths can provide you the foundation for your personality, lifelong motivations and successful high- achieving character and path. To amplify and energize these strengths of yours, engage in regular self- development and spirit enhancing exercises and activities. When you are in tune and in touch with your inner spirit, you are more connected to your soul; and this is the root and foundation of your grace and benevolence. Soul empowers you, inspires you, increases feelings of unconditional love and compassion, and motivates you to be the best version of yourself you can possibly be. Living with soul is living with heart and a sense of purpose. This is in complete harmony with your true self, when you are shining in your prime. Your sense of hope and self- honesty also come from this space and energetic vibration…

Driven/ ambitious

Another gift and strength of yours; ambition and inner drive are your fuel and your fire. Yet, like with any fire too many flames can lead to chaos and destruction, and this is something you need to be very careful with; if not at least mindful. Aggression, manipulation tactics, self- deception, deceit and a mild (or extreme) form of narcissistic personality disorder can all take over if you feel like you are not getting you way, or being recognized for your talents. Despite your strong sense of independence in most areas of life, you can become very *needy* at times. This almost always stems from some insecurity or if you are feeling out of tune with yourself. Luckily, these moments are short- lived you always get back on track, but there is still a but. Turning towards rather ugly traits and tendencies, such as spite, vindictiveness and even intentionally sabotaging or betraying others, can leave you being alienated, despised and ruining your public reputation (if you have one, which you most likely do). There are ways to exert your authority and power without squashing others in the process, and this is one of your main challenges and lessons in life. We are sorry to say, but you are prone towards psychotic and narcissistic tendencies and behaviors! There is a black and white, and a multi- colored rainbow light, to this truth.

Because you genuinely do work so hard to achieve success in life, and because you are willing to share it with others, when you feel undervalued, unappreciated or disrespected, the need to "go on the attack" is arguably justifiable; at least in your slightly clouded moments of thinking. A huge part of your personality is about being a role model or inspiration for others, and combined with your massive heart and strong sense of benevolence, you may find yourself asking questions like; *"how dare they disrespect me so deeply?" "Do I not have a right to feel this way?" "I have worked all my life to making the world a better place, so why should I allow this person to pollute my image with their illusions/ judgements/ disrespect?"* Quite simply, the basis of your reasoning and subsequent actions for your worst character traits may appear psychotic or narcissistic, but the definition of narcissism itself is rooted in a form of self- love. Despite all the negative stigma and actual "ugly" traits of narcissism, narcissists do have a strong sense of self- love. (Please note a lot of this *is* delusional and distorted. Self- love is, however, still an underlying principle.) So, as there may be a fine line within the various elements and aspects of your personality itself, working toward self- development, integration and mindful self- talk; honest self- talk and truthfulness (your Virtue); can really help you in overcoming the dark aspects of your psyche and self.

If in doubt, take things back to ground zero. Grounding yourself and focusing on your positive characteristics, such as purity of heart, gentleness and charm, grace and benevolence, enables you to re- align and re- center, further steering clear of the unhealthy side of your personality. Feeling over- worked, competitive and impatient only arise when you are out of tune and awareness with soul, therefore returning to hope and self- honesty/ truthfulness will almost certainly almost combat this. If the issues keep recurring or coming back into your life sporadically, integrate a self- healing routine into your life. Mantras and affirmations are highly effective tools to program the brain and positively influence neurons- neurological activity responsible for thoughts, beliefs, emotional patterns and behaviors.

Inspirational (A "Role Model")

Finally, you are deeply energetic, competent, charming and graceful. You have admirable self- esteem and an ambition manifesting as inspiration. You truly are dedicated to become the best you can be in this lifetime and strive for excellence. It sounds like a dream, and rightly so. Yet if you do not keep you ego in check you can often disintegrate into fearing failure and humiliation, or creating self-imposed fears and limitations surrounding your own success. You can also become exploitative, opportunistic or jealous of others' successes. Despite having it all yourself, you may sometimes wish to outshine everyone else in an unhealthy way, and further seek to "destroy" or ruin them in some way. This is you at your worst, and it can be seen as a duality to you at your best; yourself as a role model and wayshower. Your main challenge in life and golden key to your success is to define who you are, what you want to be, and why. If you are focused on your own inspirational qualities there will be no room for obsessing over anyone else's happiness or achievements. The best way to make sure you stay on track and self- aligned is to strive for balance and inner harmony in all you do.

By being too driven, too ambitious and too success- minded, you inevitably turn towards the shadow elements of your personality (deceit, vindictiveness, pride, vanity, unhealthy self- perception…). But what about your close relationships, love and friendship? What about personal passion projects and hobbies? What about spirituality or relaxation time? A shift in conscious awareness and inner focus towards all of these things could possibly be the catalyst you are seeking in life. To be a role model is to be balanced, whole and integrated; it is to have done the self-development and healing work and to further use your personally found gifts, talents and self- discoveries to be of benefit to others. "To inspire" is to stimulate, motivate and encourage- so how can you possibly do this when you yourself are not living authentically and by your truth? Your Virtue as a type 3 is *truthfulness*, and with this comes *trustworthiness.* You need to be trusted by those you wish to inspire, yet you also need to trust yourself. If you do not, you are prone to becoming too vain and deceptive just to continue to appear in a positive and favorable light; but this is a mask, a deception. To be the outstanding and unique individual you wish to be, balance your strong sense of idealism and ambitious-perfectionism with self- love and self- care.

Meditate, journal, play, laugh, love, connect on a non professional or business level, and travel often. Spending time in nature, or going on regular spiritual and healing retreats, are a sure way to soften the shadow & unhealthy aspects of your personality; forevermore.

Type 4: Holy Idea: ORIGIN… Virtue: Equanimity/ Calm. Overcoming/ softening, Passion: Envy… Fixation: Melancholy.

If you are a type 4, your strengths are as follows:-

- Self- Aware
- Expressive
- Creative
- Honest
- Inspired/ Inspiring

Self- Aware

You are self- aware, sensitive to others and to your surroundings, and intuitive. You have a gentleness and inner compassion to you which radiates out, and these are beautiful qualities to possess. Because of your extreme sensitivity and intuitive mind, a lot of your feelings and thought process are tied up in our connection with others. You put a lot of value and importance on emotional bonds and friendships or relationships, so this makes you naturally fall into lower tendencies when not at your best. Once you begin to lose touch and awareness with you Virtue, calmness and equanimity, you may fall prey into your Passion; envy. Essentially, this is due to the fact that your calmness arises from your self-awareness and ability to know what others need, and further actual care. Again- you are highly compassionate, sensitive and considerate to the welfare of others, so this in itself signifies the level of emotional support and energy you give to others. Thus, once you lose awareness with yourself your negative and unhealthy behaviors and traits kick in.

Envy arises when we are not happy with our circumstances, personal prosperity, health, relationships, self or life in general. You are very honest, dependable and responsible with a lot of common sense (when at your best), therefore you can naturally become highly critical of others. This creates resentment in the long run and can lead to minor or severe feelings of envy, or even jealousy.
Overcoming envy, your Passion, and softening this devious side of your personality

can be achieved through becoming more accepting of yours and others' imperfections or shortcomings. Learning to mediate your judgements and criticisms with a more fair and just viewpoint- one that doesn't penalize or persecute self or others and subsequently lead to feelings of inadequacy and envy- ultimately creates a level of self- forgiveness and therefore contentment. This links strongly to your main virtue equanimity. Furthermore, because of the strong emphasis you put on emotional depth and connection, another effective way you can "soften" this envious tendency is to simply learn how to have fun. *Lightening up* can work wonders and remind you that, in fact, you have a lot to be grateful for. Your motto in life to stay connected to your inner calm and associated self-awareness, compassion and sensitivities (which can be used for good), and to help let go of envy, is to accept what can't be changed, surrender to yourself whilst going easy on yourself and "lightening up," and develop and integrate the awareness to spot and speak the truth. Some things need to be said, but nothing everything does; so by following this advice you won't feel the need to always call others out- based on such a strong and rooted sense and respect for emotional melding and merging- which leads to them rejecting you and further leading to envy or jealousy. In short, go easy on yourself from time to time.

Expressive

Linked to the last point of being self- aware is that you are deeply self- expressive. You love expressing yourself through art, music, dance, love, writing, intellectual pursuits, following your goals or dreams, and often inspire others through your advanced and vast creativity. This is amplified by your intuitive and deeply sensitive nature; you can literally tune into multiple frequencies, realms of thought, emotions and higher concepts and ideals. In fact, you are so aware of feelings and inner impulses, both your own and those of others, that you may even be gifted psychically or telepathically verging on the point of being an empath. Your self- awareness and highly compassionate nature contributes significantly to your creative and self- expressive self, and whether you choose to ground this into a profession or career; you can influence others just through your thoughts, inner currents and energy alone. So what does this all mean in terms of softening the unhealthy part of your personality?

Well, to begin, being mindful of all of your strengths and favorable qualities, and further seeking to develop them and master yourself so they become integrated and embodied into your core self, will steer you away from your Passion and Ego-Fixation. It has already been expressed that you can be drawn to envy when you feel out of tune with yourself, with your Essence, but what has not yet been established is how you are very prone to *melancholy*. Melancholy is your Ego- Fixation, the aspect of yourself to turn towards when you start to feel disconnected from your true self; furthermore the other negative character traits which accompany melancholy are being extremely moody and withdrawn, and even taking on depressive tendencies. There is a dual side to this, however, as these traits always lead you onto a quest for wholeness and in turn this seeking of wholeness and harmony within can guide you on to your true path. As a type 4, your inner empathy and rather extraordinary ability to connect with others on such a deep and profound level, and feel such sincere levels of compassion for humanity and the world as a whole, can lead you onto the healing arts. In short, you may choose to become an incredible healer, shaman, spiritual teacher or wayshower, therapist or counsellor. (Alternatively this same gift and tendency may assist you to becoming a well- known and inspirational artist or creative of some sort. We explore this in the next section.)

So, with regards to the healing arts your Holy Idea of *origin* is exactly as the word itself implies- you are at your best and in your prime when connected to your essence, your inner source and your divine and original nature. As it is the moment you fall out of personal harmony and alignment with this you adopt your shadow or unhealthy traits, engaging in frequent healing and boundary-setting exercises can help you stay connected to your inner light and self. And there are so many ways to do so; journaling, writing, a number of unique therapy fields, self- healing, shamanic techniques, meditation, kundalini yoga, gentle and "inner spirit" connecting martial arts like qigong and tai chi, and virtually anything which gets you intune with your core and higher self. Also, the more you engage in such activities and stay connected to your origin, the more your Virtue will increase and the better able you are to overcome any envious or melancholic tendencies.

Creative

To keep it short, sweet and to the point, you are extremely creative. You are a natural artist and excel at expressing and connecting to the vast array of human emotions through art, dance, music, poetry- anything which allows you to express yourself. Empathy, passion and emotions become mixed in a unique trilogy where someone watching you or experiencing your self- expressions may feel like they have been witness to a divine and godly/ goddessy experience. With this high passionate connection to self, however, is bound to come its duality- and this is where problems can arise. Experiencing such intense highs means that at other moments you feel extreme lows and these lows can make you lose all faith and inspiration, even if momentarily. You start to become envious at people who are more care-free and free- spirited than you, and even sometimes jealous at how they can live life so light- heartedly without feeling such depth as you do. Of course, melancholy can also set in from your connection and sudden disconnection from the divine. So, your challenge in life and way to help transcend these depressive and jealous tendencies is to seek greater *balance.*

Balance can be attained through not taking yourself or life so seriously. It may seem that everything has to be felt and experienced with such depth and intensity, but this is just one aspect. There are 5 elements, earth, air, fire, water and ether/ spirit, and water (the element strongly associated with emotions and depth) is only one. Life itself also involves a balance of yin and yang, the dual yet complementary forces in play. Your challenge is to life with an open heart whilst incorporating joyfulness and your shadow self. Why your shadow self too and not just joy and lightness of spirit? Because, a huge and integral part of your personality and any potential path you choose involves an element of suffering, of recognizing the shadow self and melancholy, or darkness, of life. Trying to block this out, deny it or suppress it will only perpetuate and amplify it, and it will further lead to your inability to feel. A lot of what you feel and experience through your connection to the divine in self- expression outlets and forms is from a unification of body, mind and spirit; and a unification of the light and dark within. The divine incorporates both our light and our dark, and the reason you can feel and connect so deeply and in a god or goddess-like way is because you recognize the duality of life, the love and the joy and the pain and the darkness simultaneously. Your *holy origin* originates from this space of duality, acceptance and balance.

Self- acceptance is the first step and with this comes sincere feelings of inner peace, calm and equanimity. This naturally helps you overcome your unhealthy mental patterns and disruptive emotions and subsequent actions.

Honest

Your connection to the divine and your true self, through your Holy Idea and optimum form of origin, enables you to be honest and further show this honesty in times of need. In fact, your honesty can be inspirational, and no emotional display, speech or moment of real and raw vulnerability is shameful to you. This is the ultimate gift of both an artist and a healer, you appreciate and honor your emotional sensitivity and seek to utilize it in a healing or inspiring way. This in itself can help you overcome any envious feelings or thoughts you may experience, or be prone to experience. Knowing that you are helping people overcome their own problems or low moods and suffering through your gifts and talents, also opens you up to striving for inner perfection in alignment with your Holy Idea; your essence and origin (original divine nature and source). But, first you must fully develop and embody a level of honesty to allow you to do this. Seeking meaning in life through your creativity, self- expression, self- awareness and highly intuitive- sensitive- empathic nature is one route. Another is to cultivate compassion so you don't become withdrawn, over-sensitive or down, which are the direct results of feeling disempowered and disconnected from your source of spiritual or empathic/ self- aware power. Compassion recognizes both the light and the dark; it sees the dark, the "shadow" or ugly and unfavorable parts of any situation, experience or person, and aims to shed light on it. Self- compassion, therefore, is your key to overcoming the unhealthy parts of your nature.

Your key motivations in life are to surround yourself with beauty, maintain specific moods and emotional states, to retain individuality, and to withdraw when you feel particularly sensitive to protect your self- image. These require a great deal of honesty and openness, without which you may become jealous because you start to turn your attention and focus to others. Although you are not self-centered or selfish, and certainly don't lack humility or a sense of selflessness and service, a lot of your personality is focused on your own gifts, strengths and talents and how you can inspire others. With this comes an element of envy because you are always comparing yourself to others, even if subconsciously or unconsciously, and melancholy due to the inherent disappointment contained; disappointment when you start to see the world for how it truly is. I.e. not everyone is as deep or empathic as you and some people truly do not value the arts or self- expression like you do. And this is where your sense of something being missing arises and is birthed from. Yet, it is also the start to your search for higher things, inspiration and self- expression- your journey to finding and attaining wholeness comes from this harsh truth. Whichever way you choose to go is ultimately up to you, you can go up or down, towards your light or towards your melancholy and disintegration, or into your self- aware and artistically sensitive self- or self- pity, hopelessness and depression. When in doubt look towards your Virtue and aim for inner balance, acceptance and harmony, for these will see you through.

Inspired/ Inspiring

Finally, you are both inspired and inspiring and can tune into some universal archetype or image for artistic expression. You are able to transform the mundane into something extraordinary, and may possess elements of alchemy. Your ability to heighten and evolve emotions and perception or belief through emotions, fantasy, spirituality or passionate feelings and expressions is really second to none. Yet, this constant desire to stay in tune with your feelings can leave you over-analyzing everything, taking everything personally, becoming self- absorbed and being introverted, moody, pessimistic and shy or self- conscious. You have a tendency to lack all ability to be spontaneous, joyful, youthful or present, when at your worst and disconnected from your Holy Idea, and your hypersensitivity when like this may lead to severe withdrawing and isolation. This all originates from your need and desire to be inspiring 24- 7 and to be the best version of yourself, on an emotional and sensitive level. But where's the cut off and the self- love? From a different angle, it can also make you think you are different from others and that no- one else could possibly do what you do, verging on feeling superior or socially exempt from the practical responsibilities and concerns others have to deal with. This unhealthy level of apparent narcissism may appear as a form of narcissistic personality disorder and extreme egocentricity, but in reality it stems from a place of insecurity and inner pain and despair. Self- pity, hopelessness and a lack of self- worth arise when you believe you are truly "better than" everyone else, as this is a major distortion by, with and from yourself. Your true self recognizes that we are all one, arising from the same source and energy, and regardless of our individual gifts and talents and uniqueness, we are still all capable of the same expressions of talent, ability and power. In its positive expression this belief can lead to using your own gifts for good and to inspire and create; however, when muddled, clouded and deluded the negative/ unhealthy/ shadow actions and behaviors described arise.

To satisfy your inner drive of avoiding being ordinary, avoid *wounded withdrawal* and anger outbursts arising from your super- sensitive nature. You can still connect to your divinely inspired and amazing self and character without coming across as a narcissist, or withdrawing into a self- depressive bubble. Developing *authenticity* can really help with is and simultaneously align you with your true self and nature. The more honest, intuitively connected and authentic you are the better you can respond to life in a calm, balanced and emotionally- aware and present way. Finally, remember your Virtue; equanimity, and your challenge/ key: to live with an open heart integrating and balancing joy (light) and suffering (darkness/ shadow).

Type 5: Holy Idea: TRANSPARENCY… Virtue: Non-attachment/ Selflessness. Overcoming/ softening, Passion: Greed... Fixation: Stinginess.

The strengths for Type 5 are as follows:-

- Visionary
- Perceptive
- Innovative
- Self- Mastered
- Independent

Visionary

You are visionary when operating at your highest vibration. You are capable of influencing the world at large in inspiring and new ways and may make pioneering or groundbreaking discoveries or works of creation. Quite simply, you are ahead of your time and may have been called a "force to be reckoned with" when younger by teachers or elders. Your visionary qualities are you at your best, when most in tune with your essence, and when in harmony with the divine and your higher self. Yet, falling out of awareness with your favorable strengths and qualities can lead to some rather unhealthy and dark tendencies. You can become reclusive and isolated, separating yourself from reality, others and your society or community. Feelings of insecurity and instability may take hold and as a result you may become extremely eccentric, nihilistic or delirious. The opposite of your visionary qualities take over and some 5's even develop moments of horror, real fear and illusionary phobias. This is because the scope and intensity of your revolutionary ideas and extraordinary visionary creations seem truly terrifying, and you begin to feel threatened and delusional through faulty perceptions and thought processes, and self- destructive emotions. At your worst, a psychotic episode or breakdown could result and this may lead to explosive outbursts, schizophrenic tendencies and severe distortions of reality.

Learning about your Holy Idea, Virtue, virtues and positive strengths can greatly assist in overcoming these tendencies and further recognizing them before they develop. Hence, the best thing you can do is to work daily on your Holy Idea, *transparency*, and your main virtue (the Virtue), *non- attachment.* Firstly, when you become more transparent you become more honest with yourself. This in itself opens you up to clearer ways of perceiving and authentic communication. Communication is not just the direct conversations we have with others, it is also self- talk; the conversations we have with ourselves and the stories we tell. To have vision is to be able to see things from a clear, fresh, new and holistic perspective; think "big picture." Well, there is no distortion, faulty perceptions or chaotic/ psychotic thinking with the big picture.

Perceptive

You have a unique scholarly and perceptive nature to you with incredible intellectual capabilities and capacities. You thrive in wisdom and acquiring knowledge, and unlike some other personality types the need for learning and personal growth in the intellectual fields is a must for you. You may become a teacher or some technical expert in a unique field due to your keen perception and analytical skills, or you may choose a profession involving investigation or alternatively inspiring others through your other gift- vision. Furthermore, you possess the ability to detach from others or emotional pressures and dramas. This can create significant freedom in your life, but it can also create *loneliness and isolation.* Your main problem is that you are intellectually brilliant yet your close and personal relationships tend to suffer, and this is because you have a tendency to withdraw and resort back to your scholarly and intellect focused nature. This is further amplified by your basic fear, the fear of being useless, or not contributing to the world or society in some unique and extraordinary way. Learning how to participate in your own life through getting in tune with your feelings, and further by balancing your inner and outer worlds, can help you to overcome the negative and unhealthy tendencies you often resort to.

Spending copious amounts of time in solitude pursuing your own interests and re-creating yourself can help with your Virtue; selflessness- and further contribute to your ability to be transparent (your Holy Idea). However, as with everything in life there must be a balance and when you neglect other parts of your personality and basic human needs and desires, such as the ability to feel, connect with your emotions in an honest and vulnerable or transparent way, and develop close relationships and friendships, you end up closing yourself off to your strengths and gifts. The key lies in transparency, being transparent and honest with your own feelings so you can adopt and integrate the healthy parts of your personality. If you don't, you will begin to turn towards your Passion greed and Ego-Fixation stinginess- and this is due to becoming reclusive and closing yourself off. It may seem like a contradiction to your true self, as your Virtue is selflessness and being able to detach from situations, people and groups; yet, this innate drive to close yourself off inevitably creates the behavior and belief of being stingy and not wanting to give or share in your time, talents, expertise or resources. In other words, as much as your virtue is being selfless and utilizing your visionary, game- changing qualities and motivations to help others, you are still very much prone to doing the complete opposite; which of course stems from a loss of self, a loss of awareness with your true self.

Henceforth, the best thing you can do is to seek to embody your Holy Idea, transparency, so you can be authentic with your own feelings and needs, in order to be so with others and in your career or calling. Practicing non- attachment will further reconnect you to your selfless, inspiring and "desire to good in the world" nature and essence, which in turn allows you to naturally move away from your shadow or negative personality traits. Greed and stinginess only arise when you feel separated from others and the world, and discontent with your lot in life. So become conscious of your need for real, authentic and sincere connections- emotional and spiritual or soulful connections- not just intellectual connection and stimulation! In doing so, you will learn how to feel and create new opportunities in harmony with your true self and essence, so you can further re-establish yourself as the selfless, generous, intelligent and innovative person that you are. Transparency is not just being honest and authentic with others, it is about being honest and authentic with yourself first and foremost. From this space, your generous, giving and selfless feelings and motivations ripple out to meet the world; this is where you can do most good and where your Holy Idea and Virtue arise.

Innovative

You are one of the most innovative and inventive types in the Enneagram. But does this desire to always create for the benefit of others and do good in the world, or for society, lead you to become stingy and greedy? The Enneagram would suggest so. When at your lowest, you have a tendency to "hoard" which can manifest as hoarding knowledge, information, time and energy. Quite simply, contradictory to your usual giving and selfless nature (represented by your *Virtue*), you can become very greedy! A lot of this stems from a *disconnection with your body*- in other words, you lose touch with your feelings and instinctual nature and instead turn towards your mind and thoughts. But, life is a balance and we have a mind, body, emotions and spirit; we are not just the mind or mental capabilities. This ultimately means that through being out of touch and tune with your feelings, internal currents, emotions and bodily needs and functions, you start to lose touch with reality. You become purely focused in the mental body and conceptualize and fine- tune everything going on inside that brilliant brain of yours, whilst simultaneously creating (unconscious) chaos or turmoil around you. In short, you become completely immersed in your imagination and innovative qualities and intentions, that you forget your body and physical reality. This disembodiment can create problems in your perception of reality and ironically result in the one thing your higher and true self doesn't intend or want to happen; you become *stingy and selfish, greedy and cut off* from the world, and *withdraw* into your imaginary realm once again.

Fortunately, you can heal this by learning to detach from your gifts, wisdom, skills and talents, and all that you are seeking to create, and get in tune with your feelings.

Self- Mastered

With self-mastery comes shadow and, as you are one of the most self- mastered personality types, this naturally means you have some significant shadow (unhealthy) character traits to work through. You are highly skilled and analytical, perceptive and intuitive, and wise and independent. As soon as you begin to lose touch with these qualities of yours and further devolve into egotistical and lower self and mind ways, this is when and where your Passion and Ego-Fixation take over. And greed and stinginess are not just the only unhealthy displays, you also become intellectually superior, a know- it- all, unfriendly, pompous, highly judgemental or ignorant, and critical of others. You can also become emotionally superior even though it is your lack of authentic emotional bonds and connections which create the majority of these character flaws. They can almost be seen as defense mechanisms used to avoid the feelings of separation, isolation and emptiness, and to simultaneously project that image that you are omniscient and self-sufficient. This may be so, but- again- no man is an island, and we need human connection and emotional support to thrive and survive.

To stay self- mastered in your skill, whatever it may be, open up to people. Learn how to talk about your feelings and express yourself, and try and balance the importance you put on intellectual and visionary "big picture" pursuits with those you believe to be mundane or "lower" (lesser). You can still be an expert in your field and create visionary works whilst respecting your need for emotional support, love, intimacy and friendship.

Independent

Finally, you are deeply independent and utilize this to develop self- mastery, your chosen skill, your innovative and inventive qualities, and your knowledge bank or reserve. It has already been established that you suffer with forming and maintaining emotional bonds and close relationships, therefore the best way to stop yourself from disintegrating into your unhealthy tendencies of greed and stinginess is to spend sufficient time developing inner *empathy* and compassion, including *self-compassion.* Any step, action or activity you undertake to develop and strengthen these positive emotional qualities and characteristics, will further help you with any and all of your weaknesses, flaws & follies. Empathy and self- compassion should be your priority as it is from these spaces where your desires for innovation and extraordinary creation come from. Furthermore, although you don't realize it and are often unconscious of it, a lot of your independence and admiration/ appreciation/ respect for your independent qualities actually stem from your emotional connection to others. Emotions are an underlying force, shaping and creating our thoughts and mental patterns. So, your incredible mind and subsequent intellectual and visionary pursuits are actually birthed from powerful emotions and raw emotional power.

It is as if you have a shield up which completely dis- acknowledges how your emotions shape, create and influence your decisions, and this has its roots largely in the subconscious. Your *inner subconscious yearnings* are the elements shaping your choices and decisions to choose the paths you do, yet; you are almost oblivious and blind to this aspect of yourself. Why? The answer lies in your Passion and Ego-Fixation. You turn towards greed and stinginess- the unwillingness to share anything and everything you have with, well- anyone! Because your real essence and true self desires nothing more than to share, grow, inspire and connect. The apparent *disconnection* you feel is in fact your connection to others and the world around, being hidden and distorted through faulty perceptions- which arise from placing too much emphasis on the mind and intellect. Hopefully, you are starting to see the pattern and the cyclic nature of how your innovative, intellectual and mental qualities go on to create the distortions in your thinking when you disregard emotions and feelings; your emotions and feelings are an *integral* part to your strengths, talents and virtues!

Type 6: Holy Idea: FAITH… Virtue: Courage. Overcoming/ softening, Passion: Fear... Fixation: Cowardice.

If you are a type 6 these are your major strengths:-

- Loyal

- Courageous
- Responsible
- Trustworthy
- Engaging

Loyal

As a type 6 you are loyal. Your key motivations in life are to support and feel supported, obtain security and to develop significant and authentic connections. Although a mental (mind- based) type, you use your keen intellect and perceptive skills to understand the world and others, further wishing to "figure people out." You are motivated and inspired by social groups, organizations with common and shared interests, and the power of community, and you possess great faith in yourself, humanity and the world at large. Faith is your ideal, your true personality and your Holy Idea. The faith you hold can manifest as showing great courage- courage to help others in whatever way you see fit or necessary, even when you feel fear yourself. It also shows itself through your loyalty and attentiveness to others. Yet, at the same time you question and doubt yourself and oscillate between certainty and skepticism, whilst also giving in to self-created fears and a level of *cowardice*. This may appear contradictory to your confident, expressive and usually fearless nature, yet, these tendencies do still arise from time to time.

As cowardice is your Ego-Fixation- arising from a loss of awareness of faith, your Holy Idea; and fear is your Passion arising from a loss of awareness with courage, your Virtue; looking towards your strength and loyal nature can be a key way to stay connected to your Virtue and Holy Idea simultaneously. Faith arises and strengthens when you are staying true to yourself and others, and developing courage to stay strong, centered and aligned to your truth enables you to overcome fear. Connected to this is your other "shadow" traits and tendencies. Because life involves balance and often gravitating towards one extreme, you are prone to becoming suspicious, distrusting and skeptical when you lose touch with your essence. You can also use the defense mechanism of *projection* to maintain the superglue image of appearing loyal, projecting disloyalty in others when you wish to "save face," or the like. These all have their routes in fear and a temporary inability to be and feel brave in your dealings with others. Being aware of this tendency- to project your own fears and inadequacies surrounding loyalty onto others- will allow you to stay connected to your inner light, courage and faith so that you can remain strong within your path. When at your best, you are dedicated to people and projects or groups you believe in, so keep the faith and seek to develop courage in all you do.

Courageous

Your courage stems from a genuine and innate desire to use your gifts and strengths for service. Although typically a mind- based type, your courage comes from a strong need for emotional connection and support. It can also arise from need for spiritual connection and belief in the power of spirit and community. One of your coping mechanisms for falling into unhealthy shadow tendencies is to rush into action and get involved in some ideological, physical or intellectual activity. This allows you to stay in strength mode and further draw from your faith to continue on the upward and harmonious spiral. This is by no means negative, and is actually very beneficial for steering away from fear and cowardice; however, it also means that this constant need for action and engagement in something to keep you focused on your courageous nature can leave you anxious or suffering from nervous tendencies. Worry, procrastination and illusionary fears may result, or you could fall out of tune with your faith and shy away from your responsibilities.

In short, developing courage and passion for your calling allows you to remain centered, strong within and aligned to your sense of purpose, whilst simultaneously keeping you on course. Linked strongly to courage is your impeccable sense of responsibility.

Responsible

Responsibility defines you. You are practical, security- oriented, reliable and hardworking, with a highly trustworthy and common sensed nature. You can foresee problems and find solutions to them independently, yet equally work well with others and thrive in all aspects related to cooperation, harmony and teamwork. Your faith and belief in higher ideals mix well with your practical and down-to- earth nature, and anyone who knows you knows that you are trustworthy and can get a job done. However, you can also become defensive, anxious, stressed and indecisive, and at your worst can be reactive, rebellious and defiant whilst projecting your own issues onto others. This is not helpful for your lifelong goals, dreams and aspirations and can seriously damage your reputation or likeability factor. People respect you when you act with courage and passion. Furthermore, faith is not just limited or exclusive to faith for some higher vision, it also relates to faith for your every day human. Working on increasing your stability and self- reliance can help you massively in overcoming and softening your unhealthy traits. Also, suspicious and self-doubting tendencies can be eased with meditation and self- healing.

It is important that you do engage in frequent healing and self- development work. Getting in tune with spirit, or your innate spiritual essence, which connects strongly to the source of faith and inspiration; will open you to pathways for healing, self- development and personal self- expression. This will also enable you to speak your truth more freely with courage and self- assurity. In truth, compared to the other Enneatypes your problems and concerns are not many; the only things you have to be truly mindful of are becoming pessimistic, suspicious, doubtful and anxious- yet these can all be overcome and transcended through developing your Holy Idea and Virtue.

Trustworthy & Engaging

To keep it short and sweet, you are deeply trustworthy and this stems from your reliable, responsible and grounded nature. Being attentive to people's problems, needs and desires means that the people you show your support and care to trust you, they know that they can rely on you. They also know that you are sincere and this is a brilliant quality to possess. You are highly engaging, emotionally aware, intuitive, endearing, lovable and affectionate. Creating bonds and harmonious partnerships is important to you, and so is community. Anything you choose to do in the physical and material world, such as career choices, path and purpose, are all signified by your connection to others. There is nothing superficial to you, and this makes you very trustworthy and engaging. But, you still give into self- created fears and steer away from your courageous, and often inspirational, nature. One of the most effective ways to combat this is to redefine your intentions. What is the point and purpose with the hard-work you put in? Why do you do the things you do, and what are your motivations? Why do you have such a strong desire for security, close bonds, and mutually respecting loyalty?

These are just some of the questions you should be asking yourself.

By redefining your intentions and perhaps taking time daily to energize them, you are better able to stay committed to your responsibilities and personal integrity, and connected to your faith. Your faith is one of your greatest assets and it is something that will see you through the trickiest or most ambiguous of times. Working to increase equanimity, inner peace and calmness, contentedness of spirit, and emotional harmony and security- with both yourself and the world around, or at least those closest to you- will ultimately be your greatest challenge and biggest personal transformation. Once you do, however, you will find that all of your shadow, unhealthy and self- destructive thoughts, behaviors and personality traits diminish. Emotional insecurities and anxieties are actually your greatest setback to wholeness, and your key to achieving inner serenity and success.

Type 7: Holy Idea: WISDOM/ PLAN… Virtue: Sobriety. Overcoming/ softening, Passion: Gluttony... Fixation: Planning.

Your main strengths as a 7 are:-

- Adventurous/ Spontaneous
- Fun- Loving
- High- Spirited
- Quick/ Forward Thinking
- Versatile

Adventurous/ Spontaneous

You are adventure- seeking, spontaneous and optimistic. You are always seeking new experiences, connections and opportunities and often have issues with impulsiveness, restlessness and impatience. You are generally considered optimistic with a positive attitude, but this alongside all of the previous qualities mentioned can lead you to becoming obsessed with planning and gluttonous. Let's go into this in more detail.

Firstly, the egotistical tendency to become obsessed with planning arises from a loss of awareness and connection to self. Your true self- your essence- is enriched with wisdom and a "divine plan," a sense of knowing and trust in your spontaneity and inner spirit. You tend to live in the present moment and trust that your insight, intuition and innate wisdom will see you through. Referring back to gluttony, being gluttonous does not just relate to food- it also implies excessive consumption of ideas, patterns of behavior and experiences, or in your particular case; fun. Your Virtue is *sobriety*, meaning that when you are at your best version you naturally take on a sobre and level- headed state of awareness and being. Sobriety also relates to calming the mind to be still and present in the present moment, in each moment of now. Thus, learning to do this helps let go of egotistical habits and out of control desires.

This links strongly to your archetypal challenge too. Your challenge in life is to make idealism, fun and free- spiritedness practical, to ground your high- spirited and spontaneous nature in a way which supports your growth, your self- development and your mental capabilities. It is also to balance and harmonize optimistic thought with recognition and acceptance of your shadow side and self- the unhealthy parts of your personality. Gluttony comes from a space of excessive need, want and desire, which ultimately arises when you are not living in the present or being present with your body, inner source or instinctual needs. Being over-stimulated is a result of not connecting to raw, deep and authentic emotions and sensations, and this truth is reflected in your lack of emotional depth throughout life and within your interactions. Working on your emotional body and seeking to embody more emotional wisdom, intelligence and depth, therefore, can greatly assist in the softening of your *Passion* and *Ego-Fixation*, further leading to the increase in your Holy Idea.

Fun- Loving

In short, you are fun- loving, active and playful with a strong pull and desire to maintain your freedom and happiness. You have a strong aversion to missing out on experiences and opportunities and enjoy keeping yourself occupied, excited and in a constant state of flow and fun. Yet, a lot of this is to mask your inner pain and suffering or struggles. Regardless of how high- spirited, optimistic and joyful you are; *everyone* feels pain and the loss of separation, at some point, as it is a natural and innate part of the human experience. But you choose to deny and repress it, and act like you are OK when you are not. There is a huge element of *avoidance* going on in your world, type 7. You actually use this as a defense mechanism to rationalize or think your way out of feeling, and further to maintain the self image of being fine. In reality, this is done to avoid responsibility and this is a key theme with your personality type; the avoidance or dismissal of responsibility.

And this can lead to a lot of unhealthy traits and shadow tendencies. Impulsiveness, acting out in an infantile manner, addictions, offensiveness, abusiveness, manipulative and compulsive tendencies, and dealing with inner anxieties and frustrations in an immature and unhelpful way- these may all be common. The holistic and encompassing human primality of "losing control" is present in you, as you literally lose control and turn towards your Passion- gluttony. Another effect of your fun-loving nature is in your aversion to commitment, you can often be fearful, apprehensive or especially against commitment of all kinds. And this doesn't help with your Passion or Ego-Fixation! Going from one extreme to the other is common with your personality type, therefore to overcome the excessive desire to plan and create rigid structures in your life- which takes away from your adaptable, flexible and high-spirited nature- seek to balance and merge commitment into your life. Being more committed and working towards your *Holy Idea*, *wisdom* combined with a *divine plan*; trusting there is a divine flow in your life, allows you to stay versatile, connected to your inner spirit and joyful, whilst not steering towards their opposite. Also, being gluttonous is a byproduct and side effect of being uncommitted, as un-commitment ties in closely with a *lack of grounding*.

High- Spirited

We just spoke of a lack of grounding and this links to your high- spirited personality trait too. When you are uncommitted, which is one of your shadow traits, your energy is dispersed and scattered. In other words, you are *ungrounded* and this leads to a disconnection from your body, instincts and senses. It also results in you not knowing what is good for you, steering away from *sobriety*- your Virtue- and turning towards gluttony, your Passion. Being high- spirited is great and when connected to constructively and in an aligned way can create your wisdom/ plan- infused tendencies. However, being so spirited also has its negatives, grounding and a scattered & dispersed energy being one and a collection of egotistical unhealthy displays and characteristics the others. In short, you may find yourself becoming so infantile and impulsive that you suffer from extreme nervousness, anxiety and escapism. Escapist tendencies are the direct result of being so high- spirited and, sometimes, ungrounded, and further contributes to your tendency towards excessive planning and gluttony. This may manifest as food, sex, substance, shopping, event or social addiction, or anything which "feeds" one aspect of yoruself but not the others. We are holistic beings, we have a mind and thoughts, an emotional body and emotions, a physical body and impulses, instincts and physical needs, and a spirit or spiritual body with its subsequent soulful longings and higher desires. These "higher" desires include empathy, compassion and a need for deep and significant bonds and relationships.

So, through losing touch with reality from time to time, and being so up in the clouds, skies and heavens above that your energy becomes unfocused and scattered, this prevents you from knowing what is truly good for you. Your health, relationships and true path or purpose may suffer and the sobriety and level- headedness, or self- mastery, you once worked so hard to achieve becomes lost, and almost instantaneously. Fortunately, knowing this innate drive and tendency of yours, based on your personality, can help you take the steps necessary to heal and overcome it. You can soften this aspect of your nature by learning how to calm your mind and emotions, and by further finding peace and contentment in the present moment. You would benefit greatly from a daily meditation or mindfulness practice. Simultaneously, soothing sounds such as gentle jazz, acoustic or roots music, or nature sounds like bird song, whale song or rain sounds, can all cam and soothe your nerves and mind. Your spirit is a passionate one, but being grounded and down- to- earth is essential for your health, well- being and self-development.

Quick/ Forward Thinking

Being a type 7 means that your Enneatype is part of the Head Center, so in other words; the *thinking* center. There are 3 centers which we explore in the next section of this chapter. This essentially means that you are primarily cerebral, as opposed to instinctual and in the body or feeling based and in the heart. You deal with problems, issues and challenges with your mind and this is also how you like to tackle tasks and problem- solving, or finding solutions. You are also an excellent communicator when connected to your essence and are not too concerned or bothered about other people's perceptions of you, nor do you seek their approval. This adds to your quick thinking and intellectually gifted nature; being free from other people's judgements and projections provides the space you need to flourish and grow mentally. Imagining, thinking or dreaming up new ideas and future opportunities, in harmony with your spirit, comes easier to you when you are not burdened with the world's projections or external judgements. It also means you can remain free- spirited, fun- loving and joyful as your spirit will be "lighter" or not weighed down with outside perceptions. Yet, too much of a good thing may be a problem, simply because of how quickly your mind and attention can shift. You may be independent, but we are inherently social creatures, and we rely on the healthy and helpful support and influence of our friends, family, peers or loved ones. As such you find it hard to form close bonds or stay committed to certain paths and courses of action. *Depth* is an issue for you, as is staying power.

And you may have already intuited the results when this happens- gluttony and planning. Therefore, to overcome and soften this you should focus on developing depth and intimacy, whether that be platonic or not. If you can learn to "tone down" your optimistic and spirited nature you can start to shift your awareness and attention to other areas. There is more to life than fun, bouncing around on a high and always thinking about the next experience, so how about getting in tune with your inner currents, and more introspective or subconscious layers? Doing so will work wonders for your self- esteem, confidence and self- empowerment, and further allow you to overcome any nervous anxious tendencies which arise with an overactive mind. This will in turn help you overcome the need for unhealthy planning, and the turn towards gluttonous habits and behaviors.

Versatile

Finally, a major strength of yours lies in your ability to be versatile. With this naturally comes adaptability, you are incredibly flexible when it comes to life choices and decisions. But versatility for you means that you can also have so many eggs in one basket that they are smashing everywhere you walk. Furthermore, you may be holding 3- 4 baskets at once- leading to even more eggs breaking! This pretty much shows your issues in one analogy. Without wisdom, discernment and a set plan this can lead to traits and tendencies which prevent you from living your dream life. (Remember extremism- the tendency to go towards one extreme when out of touch with your essence.) The qualities you need to work on to soften your Passion and Ego-Fixation are wisdom, discernment and equanimity. Being wise and discerning combined with having a clear plan of action, in alignment and harmony with your true self, divine essence and higher self/ mind, enables you to remain versatile, adaptable and optimistic without falling prey to "lower self" (unhealthy) tendencies. This will allow you to vibrate (operate!) at your best.

Feeling genuinely grateful and appreciative for all you have and are blessed with, being joyful and blissful, feeling ecstatic and in tune with the infinite goodness of life, feeling invigorated and cheerful, resilient and zestful, being responsive to all life and sensory experiences have to offer, and assimilating all of your experiences in depth… these can all be achieved, experienced and attained with conscious self- development. Additionally you can achieve more when you are passionate and clear-headed about your life and the opportunities in it, so accomplishment and thriving intellectually- one of your natural strengths- will also be activated through self-healing and softening. Connecting to your inner divinity and being sober mentally, emotionally, physically and spiritually will enable you to achieve this. This is your key in life, and learning how to make idealism practical. In short, engage in frequent grounding activities and balance and integrate positive thinking and optimism with a recognition of your downfalls and unhealthy character traits.

Type 8: Holy Idea: TRUTH… Virtue: Innocence. Overcoming/ softening, Passion: Lust… Fixation: Vengeance.

If you are a type 8 the following are your main strengths:-

- Confident
- Strong- willed
- Powerful
- Protective

- Resourceful

Confident

Confidence is a gift of yours. You are self- empowered, inspiring and usually self- mastered, utilizing your inner strength to increase your feelings of confidence, self- worth and self- esteem. In fact, you have such a strong- willed and confident nature that you may achieve great success on the physical plane or in the material worlds. Combined with your self- confidence is your incredible assertiveness- you are not afraid to speak your truth or make your needs known. Yet, this can lead to some of your greatest downfalls if you don't keep your ego in check and continue to lose touch and awareness with your virtues and positive traits. Due to being so head- strong, confident and self- assertive, you represent and embody truth and this reflects into all of your dealings. Truth is also your *Holy Idea*, your ultimate perfection and inner essence. (Essence.) This means that you have a natural innocence to you, which is also your Virtue. Because you are confident, naturally inspiring and only wish for the best in others, this provides a strong sense of innocence in your energy, mannerism and dealings. You have no time or space for negative energy, judgements which seek to put out your light, or anything less than being the bold, empowered, inspiring and truth and justice-spreading person that you are. And this is ultimately your greatest gift; being so self-assured and head- strong- and unwavering in your intentions and beliefs- makes you wholly inspirational with a unique innocence about you. This innocence means that you face life with an open and pure heart, and further aren't susceptible to anyone else's illusions, ignorance or cynicisms. Quite simply, you embody an essence of perfection free from the "lower" qualities of human judgement and ignorance.

But, with all the light and shine must come shadow, and this is where you fall into your unhealthy traits and tendencies. Being so innocently virtuous and almost "perfect," in essence, means that when you start to lose touch with this innocence you fall into lustful tendencies. *Lust* is the polar opposite and extreme of your Virtue, innocence, and is therefore also your *Passion.* Furthermore, in your striving for truth and ultimate awareness and justice, you can often fall into egotistical tendencies. *Vengeance* is your Ego-Fixation. Operating at such a lower emotional frequency paves the way for anger, deceit, excessive force or domination, and vengeance. You can also be extremely intimidating and assert your personal confidence and authority in destructive or harmful ways. This is byproduct of lust- lust is not just sexual; it can also relate to being lustful in a non- sexual, yet specifically aggressive and overpowering, sense. This strength of yours has its many positives and steering towards your Holy Idea of truth will allow you to stay centered, morally aligned and just, as justice links closely to truth. Furthermore, being mindful of innocence, your Virtue, and finding ways to remain connected to it allows you to soften and overcome your natural tendency of falling into lustful traps and behaviors.

Strong- willed

Fairness and justice are very important to you, so much so that you often come across as intimidating, naturally superior and authoritative, and even quite scary when you exert your will. If you feel wronged or see injustice being done to someone weaker or an underdog you will literally stand up for what you believe is right. You hold considerable power to create significant and important change around you, either within your close friendship and peer groups, in your society or community, or on a global and larger scale. Your greatest virtue (and your actual Virtue) is your innocence, and this allows you to remain so strong- willed, secure and confident in your own intentions and abilities. Not only are you strong and centered but you are also enthusiastic, optimistic and joyous about your intentions and chosen actions. Nothing you do is done without any thought- there has to be some concrete and influencing motive or intended result. Your challenge in life, therefore, is to combine assertion and independence with cooperation and more "softer" approach to your interactions and goals. There is nothing wrong with being strong- willed, success- minded and goal-oriented, and in actuality these are some of your greatest strengths; however, walking over others or pushing others down in an attempt to shine or be the best only adds to your already powerful ego. The shadow and extremely unhealthy side of this is your need for vengeance when things don't go your way…

And you can literally go on the attack. You may feel like being vulnerable or appearing weak will lead to being personally attacked, or that your self- image or social status may suffer in some way. You could also believe that loosening your tight and strong- willed mental hold and control over situations, events or people will open you up to injustice, wrong-doings, or effects of literally appearing weak. But, this is not the case! There is great beauty and strength in being vulnerable, graceful or gentle, and this is exactly what you should work on if you wish to steer clear of your dark traits and unhealthy behaviors. Perhaps try incorporating meditation, mindfulness and visualizations into a daily practice and self- care routine? Meditating on grace and inner gentleness can really aid in your ability to relate to others, and further not fall into vengeful or angry tendencies. In addition, getting in tune with your inner gentleness and graceful softness will lead to your heart's true desire; truth. You can be truly admired and appreciated for all of your strengths, gifts and intentions when you learn how to soften your excessive (*lustful*) nature and balance compromise and cooperation with strong- will.

Powerful

It may come as no surprise to share that you are very, very powerful. You have a strong personal presence to you and based on your Virtue innocence, you can naturally command respect. People see you as energetic and intense and know to turn toward you for leadership. However, there is a reason why type 8 personality is also known as the "Challenger." You can be *incredibly confrontational*, so much so that you may often alienate people- even the ones who truly respect and admire you and have your back. Your innate power and natural authority means that you feel the need to control your environments, and if you don't you may suffer or be penalized in some way. Therefore, your truth- seeking and showing, innocent and open-hearted essence and true self becomes replaced with something domineering, egocentric and high- tempered. You may have a problem controlling your temper or alternatively you may resort to imposing your will on others in clearly offensive or intimidating ways… Your need to be powerful and appear strong can lead to the defense mechanism of *avoiding vulnerability* and *overlooking important emotions.* Quite simply, anything which leads to even the possibility of someone not seeing you as self- sovereign, powerful and strong with your ultimate authority intact can make you act out, and turn against yourself.

This includes excessive and lustful tendencies and repeat patterns of behavior. Being vulnerable and honest with your emotions, or your more gentler and softer side, is an instant healer. You can further connect to your true essence, your truth, power and lovingly-inspiring innocence, by working on the following and seeking to integrate these qualities into your personality; conscious action, taking initiative and self- leadership in an empowering and creation- centered way (as opposed to destructive), being a provider or protector, and embodying honor, integrity and nobleness. Also, seeking to gain power or control over someone else or some group or organization does not help your self- image! True power is using your gifts and strengths to support, uplift, empower or inspire; or to simply *share* in life's blessings and abundance. You are actually deeply generous when in tune with your true self.

So, to make sure you stay aligned to your Virtue, essence and innate holiness or divinity, take steps to become more conscious and grounded in your own strengths. This will allow you to step into true leadership and positions of power and authority, so you won't feel the need to gravitate towards egocentric or destructive tendencies. Also, as a very yang and masculine personality type, have you considered developing and integrating your yin?!

Protective

Despite your tendencies towards aggression, anger and dominance, you are fiercely protective. You always support the underdog and have no fear storming into a crowd of people with your head held high and personal integrity and morality soaring, to protect someone in need. You are decisive, authoritative and commanding- a champion for the people and someone others can look up to. You have honor, incredible morals, integrity and enthusiasm and, when at your best, use your strength and providing nature to become the hero, or at the very least generous and magnanimous. So, what happens when you fall out of tune with with these admirable qualities? Well, in short, you turn against the people you have sworn to protect and stand up for. Self- mastery turns to self- pity or self- doubt and you forget the whole purpose and intention of your actions. You may also be prone to becoming manipulative, mainly due to the fact that you feel you have an intrinsic connection and "right" to twisting or distorting the truth because everything you do and work towards is to inspire and help them. In other words, you believe you have a natural hold over them for the love, grace, assistance and strength you put out. Of course, no- one likes a savior or a martyr and this can backfire on you. It is further ironic that you are prone to this type of habitual behavior due to having an underlying fear of being.controlled by others yourself.

Tapping into your protective, supporting, empowering and providing qualities can help you to lose the need for vengeance or to simply turn against the people and groups you genuinely wish to serve. Looking towards the energetic associations and qualities of personality type 2 can also be extremely beneficial to your innate tendencies. Type 2 embodies the energy of a helper and healer. They are empathic, service- oriented, friendly and warm- hearted with gifts for compassion and deep understanding. Becoming more like 2's in your mannerisms and attitude can *amplify* your strength and natural desire to protect and serve. This can also help you to appear less like a tyrant or dictator, which are other shadow aspects of your personality, and more like an inspiring role- model or leader.

Resourceful

You are incredibly resourceful, confident and courageous. You are not afraid to go after your dreams, break the mold or dream big; success, prosperity, abundance and ambitious goals & aspirations are no strangers to you. You could very well be in store for historical greatness such as through creating lasting legacies, or following a path of destiny and conscious action. Self- mastery is not unheard of in type 8's, nor is living a life of service to a higher power or divine reality. You have "vision" and the practical skills and gifts to accomplish (manifest) it.

But why do you become vengeful or lustful and give in to your Passion and Ego-Fixation? Let's answer this. One of your driving fears and basic survival flaws is the fear of being harmed or controlled by others. You want so desperately to be self- reliant, self- autonomous and self- empowered, yet this can drive you towards the things you fear most. Or, you often start to embody the traits you so strongly fear or despise. Your basic, foundational desire is to protect yourself, the life you have created and your persona; the image you project out into the world, and further prove your strength and resilience to others. With this comes your self- emphasis on resources and their importance in your life. You have a tendency to place a lot of significance on resources, material prosperity and abundance, and physical success- how much has someone achieved and how many accomplishments do they hold? This is how you define power and strength. But you are also sincerely generous and giving, and your innate drive is centered around innocence; the genuine state of being and emotional frequency of interacting with others and the world in a loving, harmonizing and pure way. In short, you are pure, sincere and vastly heart- centered, which means you live with an open and honest heart.

And it is this exact gift of yours which can result in your shadow and unhealthy tendencies. To you, everything you stand for and represent is your strength and if anyone ever saw you as vulnerable or anything less than wholly strong, you would appear weak. Weakness is an *absolute* no- no for you, as it symbolizes everything you seek to master and overcome. Being so loving, giving and pure in your dealings needs to be met with honesty and truth, and people on your wavelength who see, understand and respect you, otherwise this will question everything you stand for and believe in. It would also question your honor and integrity, and the strength of your beautiful (but secretly vulnerable) heart. So, if or when any human shows any sign of disrespect, unappreciation to your gifts and kindness, or "rejects" you in any way, shape or form; you may subconsciously go on the attack. This is purely a defense mechanism masking the fact that you are extremely sensitive and fragile. Any displays of aggression, excessive dominance, force or intimidation are your way of protecting yourself and your beliefs, which are integral to your personality and entire identity.

So how can you stand up for yourself and what you need, be resourceful and inspiring, and stay strong and connected to your truth, without coming across as a bully or tyrant? Can you prove your strength and retain your innocence without losing a part of yourself in the process, unconsciously steering away from your essence? Fortunately you can and there are many ways to do so. Lust, either sexual or philosophical, can be overcome with care, compassion and fine- tuning your sensitivities, and vengeance or the confrontation and anger you give can be softened with; 1- working on the way you communicate, and 2- empathy. Empathy is your superpower and secret shield and *mindful empathic communication* can work wonders for your self- confidence and self-esteem. By getting to the root of your buried insecurities, gracefulness, gentleness and healthy vulnerability can begin to become integrated into your character and persona. This will only empower and strengthen you further to support and inspire others, never losing touch or awareness with self, and self-mastering yourself even more profoundly. Success is yours for the achieving.

Type 9: Holy Idea: LOVE… Virtue: Action. Overcoming/ softening, Passion: Slothfulness... Fixation: Indolence/ Laziness.

Type 9's strengths include being:-

- Peacemaking
- Mediating
- Creative
- Receptive
- Harmonizing & balancing

Peacemaking

You are a peacemaker at heart and live for love. Love is not just romantic, it can also be universal, platonic and the love you feel for your fellow human or for the planet and animals. Your Holy Idea is love, and this makes you incredibly compassionate and empathic. When at your best you are at one with yourself and the world, autonomous, self- sovereign, connected, empathically aware, present and peaceful, with an inner contentment and equanimity. You are also highly receptive, accepting, emotionally stable and secure, trusting and patient with self and others. You are generally good- natured, positive and at ease with yourself, and you certainly don't suffer from self- consciousness or pretentiousness. All of these qualities inspire you into action, your Virtue; and further inspire your Virtue. Action is your will- power, your purpose and your intentions- it is the energy of your true essence and self and your motivations in life. There is a reason why your type is literally known as the Peacemaker, you seek to bring peace and harmony to your environments and the people in your life. This is you at your prime; love- fuelled and action- oriented, influential and even inspirational.

Yet, you have a tendency to fall into inertia, slothfulness (your Passion) and laziness (your Ego-Fixation) when you become out of tune and touch with your essence. You are also a "body center" type, meaning that you are primarily in tune with your instincts, gut and inner feelings (as opposed to being in your thinking or emotional centers). This signifies that you are deeply instinctual and anything which makes you lose touch with your instincts sends you into states out of tune and harmony with your true self, i.e. slothfulness and laziness. Due to such a peace-loving nature, you additionally have a tendency towards narcotization which is the habit of dulling your senses and awareness to avoid confrontation or conflict. You also may do this to maintain and project the self- image of being harmonious and content 24- 7, even when you are not. This keeps your persona in tact but leads to built up disharmony and discontentedness, which, in the long run, can lead to very occasional and angry outbursts. Narcotization can include food, drink, drugs, t.v., or simply repeat patterns of thinking or habitual responses to keep you in a state of reduced awareness and feeling. In your desire for love and harmony you "block out" things which may bring disharmony, confrontation or minor temporary chaos. To overcome all of these unhealthy tendencies, work on developing self- love and integrating a self-care routine. Self- love and self- care can assist you in staying on course dedicated to any goal or service- oriented dream you may hold.

Mediating

In addition to your peaceful and peace- loving nature, you are also extremely mediating. You have a natural tendency towards diplomacy and taking on supportive, caring and harmonizing roles; you can bring others together through your intentions, energy and often unwavering faith in unconditional love and compassion. Conflict is not something you accept or tolerate if it doesn't need to happen. Yet, because of such an extreme aversion to conflict or confrontation of any sort, which often leads you into conscious action, this innate drive of yours can steer you into slothfulness and laziness when you become out of balance. Furthermore, because being so mediating means that you also *merge* with others, i.e. you frequently lack personal boundaries and blend to your immediate environment, people or situations (which is further amplified due to being so receptive, compromising and *all- embracing*); this leads to inertia and physical laziness. In an attempt to avoid conflict or tension altogether, you appease those around you and suffer silently yourself. This suffering may not be mental or emotional- you actually take on a very docile and compliant, yet peaceful, state- however, physically this is not good for your health. Many type 9's suffer from lower back problems, a space where stored tensions and repressed emotions often accumulate.

So, you discharge which is also disconnecting from your light, willingness to engage in healing and creation- centered conscious action, and ability to feel on a deep and authentic level, in order to maintain a "safe" equilibrium. This safety net you subconsciously or unconsciously create for yourself is not actually that safe, as it opens you up to other people's illusions, ego-projections, distortions and judgements; which in turn cause health problems. Physical tension and stored stressed can result as mentioned, and your strong tendency to merge and blend further makes you sacrifice core and essential aspects of yourself. Love- your Holy Idea- almost become lost, and usually to lesser realities, and your Virtue of inspired or divinely guided action dillutes into a reality where you become completely complacent or compliant- one of two extremes. Why complacent, you may be wondering? Because this is another shadow (unhealthy) aspect of your personality. In your desire towards right action you may sometimes momentarily take on an alter- ego where you feel smug, self- righteous or overly proud of your whole demeanour and intentions/ actions towards love and harmony. And this is largely due to a self-misperception and distortion in your way of thinking based on the accumulated tensions resulting from your compliance and appeasement.

Therefore, working on your need to almost-sacrifice yourself, your truth and the love and sincerity you hold can help overcome many problems. Seeking to develop *stronger boundaries* is one profound and powerful way to do this, and in fact can be your saving grace for virtually all of the issues and unhealthy traits & tendencies your personality type suffers. Personal boundaries can be developed with meditation, self- healing and spiritual development.

Creative

Due to your ability to merge with others, environments and unseen worlds, you can be very creative. Your highly receptive nature means that you can literally merge and bond with abstract ideas, concepts and archetypes of consciousness, and your love of peace, harmony and cooperation supports this. It is your creative energy and spirit which may lead to a lot of the tendencies and desires you hold for inspiring others or taking on a supportive and caring role. You are optimistic, supportive, reassuring, synthesizing and communicative, and further have a deeply calming and healing influence. Anything you choose to create or whichever outlet you decide to express and channel these innate gifts of yours into can result in great success, or at the very least extraordinary works of creation. Yet, again, your pull to keep the peace and evade certain issues or apparent problems gravitates you into falling into laziness or slothfulness. How can someone so creative, optimistic, inspiring and energetic fall so easily into inertia and depression? Because this is an integral element to your personality type when you lose touch with your true essence and self- depressive- like tendencies. In your sincere yet extreme pull towards mediation, diplomacy and peace you can often take on pain, suffering and sadness; some of which is directly yours but other parts of which are not. This desire to minimize or completely reject anything upsetting or un-unifying (disconnecting) creates the rejection of parts of your own self and reality, and this essentially dampens your spirit and your light.

Your creative abilities and susceptibilities to artistic subtle influences diminish, and your imagination becomes squashed with anxieties or fears of separation.

And this is your deepest and most ultimate fear- the fear of loss or separation. You thrive in connection, authentic bonds, emotionally charged and real friendships and relationships, and human affection exchanges; anything which seeks to threaten this puts you in your idle, lazy and sloth- like mode. So, moving forward and to heal and soften this innate tendency of yours, learn to be more trusting of others and yourself. Recognize that life involves separation, pain and suffering, for it is part of the human experience. Come to terms with it and practice self- acceptance daily. Also, seek to integrate and harmonize your light with your shadow self, as this will enable you to accept loss and separation better. This in itself will allow you to be your best self, all- embracing, accepting, peacemaking and creatively brilliant; further connecting you to your healing nature so you can heal conflicts and step into the meditating and diplomatic role you love so much.

Receptive

Your receptivity can make you extremely sensitive but this is where you thrive. Both spiritually and creatively- you can become a shining light for others. Although you are a body based type, you are also incredibly empathic, sensitive to subtle energy and compassion, and from another perspective these qualities of yours can be seen to arise from your instinctual nature. Your instincts are fine- tuned and you are also deeply connected to your intuition. Your intuition is your guiding light, it is the part of self which "knows" things or senses things without being told. It can help you make the right choices in life, guide you down the right paths or inspire you into action. This is because your intuition is closely connected to your higher self, your higher mind which is in connection with the divine, a universal consciousness or higher perception. Being receptive signifies that you are in tune with subtle and/ or spiritual energy and the more subtle parts of yourself. You may have profound dreams, naturally be able to lucid dream or astral travel, or you may be very connected to the subconscious realms where universal or collective archetypes, imagery and archetypal concepts are rich. Many 9's have an intrinsic link to these avenues for healing and self- exploration, as your Holy Idea is love, or unconditional love. Love is the strongest and most powerful vibration and force in the universe!

By coming to terms with your weaknesses, which are a direct result of losing touch with your Virtue, gifts and Holy Idea, you can start to make the necessary changes for you to lead your best life. For you, Buddhist philosophy and principles can greatly assist with your self-development and healing journey. Buddhism teaches the principles of impermanence, interconnectedness, non- attachment and that all life involves suffering. Well, considering your true nature and subtle fuelling drives; these are perfect for you to soften the unhealthy parts of your personality. You can remain inspired into loving and harmonious action through accepting both your own and the collective shadow and coming to terms with the pain and loss, or disconnection and separation, inherent within life. This way, your realizations can inspire and motivate you towards your Virtue- action- and further towards staying aligned, centered and committed to your beliefs, intentions and inner strengths.

Harmonizing & Balancing

You are driven by an underlying desire to create harmony in your internal and external worlds. This is your key motivation in life and your challenge simultaneously. When expressed and channeled positively, you are a gem and true rock for those you love, whether that be in your close circle or friendship group, or for family or colleagues. You have a gift for holding others together; yet, due to being so balancing and compromising you are also deeply adaptable. Although this does have its positives, being so flexibly minded means that you often fall into repetitive patterns of losing your way. You may forget yourself or become absorbed into lesser or lower realities, such as those of toxic, ill-intended, narcissistic or simply unsavory character people. This can lead to severe indecisiveness on a minor level and complete harmful changes on a major one. You may fall into the wrong crowd, lose total sight of your ambitions and dreams, or misalign completely from your talents, goals and innate gifts. Of course, when expressed constructively this allows you to be the mediator, peacemaker and harmonizer in any situation, but it is important that you first find ways to protect and ground (center and align) yourself, and further integrate these self- help methods into daily life. In short, your sanity, well-being and connection to your true self and essence all comes with healthy boundaries.

Finally one of the most effective and beneficial things you can do is to integrate body- based activities into your life and any daily routine. Walking, dancing, 5 rhythms, yoga, exercise, tai chi, creative cooking, gardening, diy- these are all perfect for your personality. Also, find activities that aren't necessarily active- oriented but do stimulate your senses. Music, arts and crafts or a creative hobby or passion pursuit are all fine examples. It is highly important that you learn how to balance and unify conflict and the "darker" or shadow (less favored/ undesirable) aspects of life, and merge them with acceptance, compassion and understanding. You don't have to like them or advocate them, but you should accept them so that you can alchemize the lesson and pain for inspiration, helping others or simply being a gem in someone's life. Also, it will stop you from falling into disintegration towards slothfulness and laziness.

Balancing Your 3 Main Centres

The Enneagram teaches that there are three main centres; the Head, the Heart and the Gut.

The Head is where we process information, it has intrinsic links to the mind and mental process, or intellect, and relates to Personality types 5, 6 and 7. Problem- solving, thinking, imagination, planning, memory, intellectual pursuits, perception and information-gathering all come under the Head centre's realm. The Head is the *thinking* centre.

The Heart is where we process information through feelings and relates to relationships with self and others, self- expression, emotions and connections. Personality types 2, 3 and 4 tie into the Heart centre. The Heart is the *feeling* centre.

The Body or Gut links to our intuition, instinctual responses and gut- feelings. Personality types 8, 9 and 1 relate to the Body centre. The Body is the *instinctive* centre.

One of the main purposes in addition to learning about your Enneatype, Holy Idea, Virtue, Passion, Ego- Fixation, and strengths and weaknesses is to balance these three centres within. Although each centre relates to

3 specific personality types, this does not mean balance and harmony cannot be attained within. In other words, you can learn a great deal about yourself through exploring the other Enneatypes in relation to their corresponding "centre."

Before looking at how to balance them, let's briefly explore each Personality type in relation to their corresponding centre.

Type 8: 8's respond to triggers and emotional issues through anger and in some physical-instinctual way, such as raising their voice, becoming more aggressive and domineering in mannerism, and seeking to exert control and physical authority over others. Lust, their Passion, is a physical quality and vengeance, the Ego- Fixation, is also manifested and represented as a physical act.

Type 9: 9's also have very instinctual Passions and Ego- Fixations. Slothfulness and laziness/ indolence are both rooted in disconnection from one's divine physical self, and accompanying characteristics and tendencies include denying or repressing anger and natural instinctual shadow traits. In the pursuit of harmony and idealism 9's will deny their instinctual urges and turn away from their whole, balanced and unified self.

Type 1: 1's equally have primarily instinctual responses and inner drives when they lose touch with their true essence. The Ego-Fixation of resentment and the Passion of anger come into play when they are not having their physical needs met, and this further creates tension on a physical level. 1's also attempt to control or repress their anger and inner gut- feelings.

Type 2: 2's display Passion and Ego-Fixation in harmony with their feelings, thus showing how they are ruled by their corresponding centre. Pride is a representation of losing touch with their strengths regarding their connection to others, and this is often expressed through wanting or getting people to like them. In short, 2's rely on positive emotional experiences and feelings with others. The same as shared for pride is true with flattery.

Type 3: Again, just like type 2, type 3's are primarily concerned with their connection and interaction with others. Deceit and vanity, the Passion and Ego-Fixation respectively, are birthed from a disconnection with self and their true self, which is always in full force when they have authentic bonds. If 3's are out of tune with their feelings and sincere, authentic and truthful emotions, their virtues are diminished.

Type 4: The envy and melancholy felt by 4's when they lose touch with their essence, Holy Idea and Virtue represent a basic need to feel and be in unity with others. Type 4's are concerned with their individuality, creativity and gifts and talents, and how they can tune into these to inspire or experience deep connections with others. Feeling and authentic human bonds are the keys here.

Type 5: 5's are part of the thinking centre and this is clearly reflected in their traits and tendencies. Avarice (the Passion) and stinginess (the Ego-Fixation) both stem form a mental distortion or imbalance which desires excessive material wealth and prosperity; a lot of desires in the physical realm are birthed from the mind. 5's at their worst become secretive, isolated and emotionally and mentally withdrawn, and at their best are scholarly and perceptive.

Type 6: 6's negative/ shadow attributes include fear, cowardice and pessimism, and these all begin in the mind, or in some mental distortion. In their positive self they are problem- solvers, attentive and philosophical-again all mind based. Losing touch with their true self can make them doubtful and anxious.

Type 7: Type 7's turn to gluttony (Passion) and planning (Ego-Fixation). The need to constantly plan and forward- think is due to a disconnection with the present moment, with peace of mind in the now; and gluttony is does not just relate to food but also an excessive accumulation of beliefs, ideas and perspectives. Sobriety, their Holy Idea, requires calmness of mind and becoming conscious in the present moment.

So, in terms of self- discovery and personal transformation, it may be clear how you can learn about each Personality type to help balance your own inner currents. Just because you resonate with one Personality type it doesn't mean you won't have aspects from the others. Remember that the Ennegaram is only one system of self-discovery and analysis; we also have astrology and other esoteric systems playing integral significance in our personality's make-up. Numerology, for example, shows us how our date of birth and name can be highly important to our growth and self- awareness. In astrology, it is not just our sun sign which affects us deeply but also rising/ ascendent, moon sign and all other planetary placements. Quite simply, the

Enneagram may be a powerful system, yet it is just one system.

Furthermore, connected to the three centres as expressed in the Enneagram as a tool for spiritual and empathic development, is the reality of the chakra system and how closely connected these teachings are to the Head, Heart and Gut. The chakras are our bodies energetic counterparts, energetic portals of energy responsible for health and well- being on a physical level. If there is a block, imbalance or disunity in a chakra and among all 7 (we have 7 major chakras), this can manifest as a variety of mental, emotional, physical and/or spiritual health concerns. Chakra literally means "energy wheel" and comes from ancient Sanskrit.

The Gut/ Instinctive Centre, In Relation to the Sacral Chakra

The *sacral chakra* is the energetic wheel or portal corresponding to the Gut, or Instinctive and Body centre. The sacral relates to emotions, creativity and sexuality- it can be seen as a trinity and merging of all three. Repressed emotions and past painful memories in friendships, sexual bonds or relationships often store in the sacral chakra.

If there is a block in this centre's energy portal, this can cause major issues in relationships and creative expression, and in the way one connects and relates to others. Emotional blocks, imbalances and disharmonious are often manifested in the sacral chakra.

In relation to this same spot being the gut and instinctive centre in the Enneagram, the same is true with the chakra system. The sacral is also known as your "gut- centre" as it is where instinctual and primal urges and desires come from. We are both spirit and human, therefore we have an animalistic essence to our nature. Primality and passion are two fundamental elements of life, and if we begin to repress or deny them this where problems can arise. It is interesting to know that a lot of issues with the shadow self and personality take root in some self- denial, largely repressed or rejected part of self, and this usually to almost has its roots in the sacral chakra's energetic associations. Desire, lust, a need for intimacy and connection, and deep emotional longings all link to the sacral and the shadow self equally. When we look at the Enneagram and its teachings, the shadow self (shadow or dark personality traits) and the divine are both integral to Enneagram philosophy. Furthermore, instinctual responses and needs are birthed from our sexuality and desire for human connection, so are you starting to see a pattern here?

Learning about the sacral chakra can teach us a lot in terms of understanding the nature of the Enneagram and the Gut/ Body centre's Personality types specifically. Finally, when the sacral chakra is blocked, i.e. there is no longer a free flow of energy from or through it, this can close one off from their intuition. In chakra philosophy it is believed and taught that the sacral chakra is closely linked to the Third Eye chakra; the seat of vision, psychic ability, perception, connection to the divine and the higher self, and intuition. Instinct and intuition are not the same, but they are connected. This "natural cord" or link between sacral (lower self/ primality/ emotions/ sex centre) and the third eye (higher self/ centre for intuition and psychic-spiritual gifts/ perception) provides key insight into how we can balance our Gut and Head centres when using the Enneagram. Quite simply, learning of the energetic associations of these chakras opens up new portals to learning, self- discovery, understanding and transformation.

The Heart, In Relation to the Heart Chakra (Main Centre for Empathy, Love and Relationships)

The heart chakra is actually known as the central chakra is this unique system and this is due to the heart being directly between the higher centres/ charas and the lower ones. There are believed to be 7 main chakras, or energy portals, on the human body and these are connected by the heart. The heart chakra symbolizes empathy, unconditional love, caring, kindness, relationships, friendships, a balance between dependency and independency, and a respect and care for nature and the planet/ environment/ natural world. The heart chakra connects "lower" feelings and emotions, including our basic instinctual urges and desires, and "higher" ones. (We go into this in more detail in the next section!)

The main word and quality associated with the heart chakra is empathy, and as we have already written a bit about this in *How to Use and Connect to the Holy Idea and Virtue to Soften the Unhealthy Side of Your Personality Type, "Type 2,"* we won't go over this again. But it is important to understand the importance of empathy here and how it relates to both the heart chakra and Heart centre in the

Enneagram. Once you start to connect to your own inner divinity, internal currents and true soul's purpose, you begin to feel more in tune with yourself. (Your self.) Well, with this naturally comes feeling your inner energy currents and being more open to spirit and subtle energy.

The Head, In relation to the Third Eye and Crown Chakras

So, your Head centre is naturally connected to the third eye and crown chakras. Your third eye as briefly discussed, is your psychic centre. An open and healthy third eye chakra can lead to the ability to perceive subtle energy, enhanced and evolved intuition, spiritual gifts, access to dream states, intellect, innovation, original thinking, and imaginative qualities and thought process. As it is located just below the crown chakra, it also assists in the free flow of energy from the crown to the root chakra; the base chakra relating to security, grounding and feeling connected to both your own body and the earth. As for your crown chakra specifically, the crown is at the top of your head and is known as the seat of cosmic consciousness. Associations include universal truth and wisdom, higher perspectives and understanding, spiritual awareness, mediumship and channeling, and a connection to source/ the divine/ some higher power. The crown chakra, however, is not exclusive to all things supernatural or spiritual, it is also very much connected to the earth. This is because the energy which flows through all 7 chakras is known as kundalini, or shakti, when awakened, in healthy flow and harmonious, and this flows from crown (all the way at the top) to the root (all the way at the base of the spine, by the pelvic region and reproductive organs).

Why is this necessary to know, you may be wondering? Well, just go back and read *Enneagram History, Origins & Influences* in chapter 1. A connection to the divine and *essence* is inherent within many ancient and spiritual traditions and teachings. Your kundalini energy is also known as "serpent power" and has strong ties to the story of Christianity, regarding the snake in the Garden of Eden. The snake as a symbol essentially represents our sexuality, inner divinity and whole, unified and balanced state of being. There is a reason why those may choose to repress or deny their own primal, sensual and sexual urges and nature call this "evil." Dark and light, evil and good, and shadow and dark; the principle is universal and it is furthermore contained in the cells of our very own DNA. Meditation, sound healing, spiritual exercises and self- healing can all help us access the core of our beings and our ancient memories simultaneously, and assist us in understanding our own chakras- as energetic portals stemming back to ancient Sanskrit times.

So regardless of how deep you wish to go into the chakra system and how you choose to integrate it into your understanding and application of the Enneagram, having a basic knowledge and awareness of the three chakras; the *Sacral*, *Heart* and *Crown*, at the very least; can open new channels for learning and self- discovery, also furthering your understanding of the 3 centres and how to balance them.

Balancing Your 3 Centres Within

Gut/ Instinct (body), Heart (feeling) and Head (thinking) are within each of us regardless of our Personality type. It is advocated that whichever centre our Enneatype resides in is the centre we are *least* able to function, because the psyche naturally has some ego blocks and distortions. For example, type 9 resides in the Body centre, yet type 9's struggle most with feeling connected to their bodies and often lose touch with their inner vitality and physical essence. Integrating all three centres, therefore, is a step towards self- awareness, growth and shifts for change.

To balance the Body centre….

Frequently engage in physical movement. Physical movement releases trapped emotions and memories and can free up energy pathways for healing and growth. A lot of what is stored is repressed, unconscious or rejected; either from being too painful and difficult to deal with, or through some self-denied shadow trait or personality aspect. Movement which is fun, joyous and connects you to your whole self can allow for great release, letting go and transformation. Also, anything which gets you in tune with your senses. Dance, tai chi, martial arts, sports, walks in nature, kundalini yoga, swimming, making love… all of these can work wonders for your gut and instinctual energies.

When you engage in any physical exercise or activity that connects you to your body, to your senses, stored and often trapped emotions, trauma and wounds become loosened. We humans also tend to store a lot of negative or harmful beliefs which we are completely unconscious of. We may be holding on to things which do not belong to us, keeping us out of alignment and entwined in beliefs, realities, stories or "frequencies" which are harmful to self. Anything which brings us back into our body and therefore back into our instinctual essence and "gut," therefore, can bring great inner balance and harmony. Due to the fact that emotions and higher realms and channels of thought and feeling, such as the intellect, our intuition and higher self holistically, are deeply connected to our lower self (gut/ stomach/ primal instincts); this naturally means balance is activated, and the body's self- healing mechanisms initiated.

To balance the Heart centre….

To heal your heart, engage in daily affirmations, mantras or self- healing activities and exercises. Any sort of therapy will help greatly and meditation, learning a Healing Art, receiving energy healing, paying attention to your dreams, or receiving guidance and counselling can all assist you. One of the best ways to balance your Heart centre is to spend time in nature. Connecting to the elements creates powerful feelings of grounding and security and feeling at one with the world, and yourself. Elemental energies also assist in

balancing your emotions, inner mental thought processes- regardless of how chaotic they may be, feeling at ease and at peace with yourself, and increasing vitality. A lot of imaginative and artistic self- expression and ideas can arise from being in nature, becoming an observer of your thoughts and emotions and subsequently filling yourself with emptiness. Nature increases our flow of chi, the universal life force energy responsible for health, vitality and well- being- and our spiritual awareness. (*Chi* is the force many ancient kung masters and martial artists work with to develop their internal strength, core being and inner spirit!)

Finally, you can cultivate empathy or seek to embody mindful empathic communication, which can be extremely catalytic to balancing your Heart centre. Communication is the foundation of all relationships, soul ties and bonds; without communication- either verbal or through our intentions and subtle impressions- we would not be who we are. Communication defines us; it is essentially us. With mindfulness and empathy combined any lingering issues in your emotions, interactions or intentions towards others can be healed and harmonized. You can further start to accept the parts of yourself you may have blocks or feelings of shame about. In short, mindful empathic communication is a key ingredient to your well- being success.

To balance the Head centre…

Intellect, reason, rationality and logic can all be connected to and developed for balancing the Head centre, as these are the key qualities which lead to its balance. Intellect, planning, mental organization, problem- solving… any and all of these should be cultivated through mindful meditation. Mindful meditation is where you meditate with a specific focus and intention, for example if you choose to meditate on cultivating problem- solving skills you would set an intention and use a combination of breathing and visualization techniques. Meditation is simply contemplating and "going within" to increase self- awareness, and bring an expanded state of awareness and presence into daily life. You can engage in it to develop mindfulness, become the observer of your thought processes and emotions, and expand spiritual awareness. When used with visualization, you can further visualize the quality or skill you wish to embody and enhance and use natural laws like the Law of Attraction (LOA) to amplify your manifesting abilities. The LOA works on the principle of frequency and vibration- everything in this physical universe holds a specific pattern or encoding of information and through techniques like visualization and intention setting, you essentially utilize your thoughts and mental power to "project" an intended vision for future manifestation. This mental projection influences reality through the power of the subtle energy in place.

In addition to these, your Head centre can be balanced through working to connect to your third eye and crown chakras and further integrate the qualities associated. Writing is also a highly effective method to express your thoughts and expand your intellect, whilst simultaneously sparking your memories to help you in other areas. You may choose to educate yourself on the Enneagram, or at least your own Personality type, and write down all you can remember. Alternatively simply copying the wisdom and information contained can assist in activating your mental capabilities and taking you out of your body or feelings, and in to your head. If you feel yourself to be too emotional or "watery," a particular emotional quality, then why not work to increase your inner fire or inner air? The air element is associated with communication, the mind and all sorts of mental abilities, including the ones expressed above. Air can be cultivated through spending time in nature, specifically in the fresh air, and watching birds or working with divination; the divinatory arts. It may sound a bit "woo" or "out- there," but have you tried trying to make a feather levitate through the power of your mind?! Regardless of whether you can actually do it or not, the sheer intent and mental power and capacity for belief alone will trigger your neurons in new and transformational ways.

Transitioning from "Asleep" to "Awake" - A Spiritual Exploration

You can use the Enneagram to transition from asleep to awake. What does this mean? It means that in modern society there are endless distractions, intoxicants and pollutants which can cloud our minds and cut us off from a spiritually connected reality. We may lose awareness through drugs, substances, food, technological pollution, unhealthy beliefs or repetitive destructive thought patterns or emotional cycles. We may unconsciously choose to reside in a limiting and 3- dimensional reality, unaware of life being a *multi- dimensional* experience and the subsequent joys, wonders and new heights and levels of awareness, and experience, this brings. Quite simply we often walk around our lives asleep, without mindful thought and feeling and lacking conscious awareness; awareness of a higher or divine reality.

Learning about your Personality type can be a spiritual awakening in itself. Knowledge is power, and wisdom allows you to *know thyself*. When reading about your strengths, weaknesses, gifts, talents, flaws and shadow personality traits and aspects, your brain becomes activated or sparked in new ways. Memories long forgotten can come into your conscious awareness and it may appear like a torch is shining into dark and neglected spots, of both your subconscious and unconscious mind. The mind, self and psyche and extraordinary things, in essence; and this is a huge element to the Enneagram and the wisdom inherent. *Essence*. Waking up to yourself is waking up to your essence, your innermost true self, true core and soul. We all have a soul as we are timeless and infinite beings, and- again- life is a multi- dimensional experience. Time is also non-linear and there is a spirit which flows through all living things, through the universe, through ourselves and through the natural objects, entities and forces of the world. Our physical bodies are channels- vessels waiting to be filled with consciousness. *Consciousness and energy* are the essence of life.

So, transitioning from "asleep" to "awake" is like waking up from a dream, and in this dream our senses have been trapped and suppressed. How could we possibly know we have shadow traits, such as a natural Passion and Ego- Fixation, when we don't *know* we have these? If we have not been taught about ourselves, the real stuff; our inner motivations, desires, wants, needs, strengths and follies; how are we to understand ourselves, and furthermore, how are we supposed to grow and change? We can't grow and change, and this is a major issue with society and further why subjects like the Enneagram, Numerology, Astrology and Metaphysics are still deemed "esoteric." Society doesn't teach us about the soul, our individual and specific shadow aspects from various angles and schools of thought (ones which directly and actively help us!), or anything non- curricular. If we want to know about these things we need to learn them ourselves, yet these topics aren't made available or known to the average 16 year old, or even 21 year old for that matter. It is not until we decide to travel and explore other cultures and countries, or take the leap to go on a spiritual retreat and visit an ashram, monastery or Vipassana, that we only start to see how much wisdom, knowledge and teachings are out there. Meditation and spirituality are only just starting to become adopted by Western society, but they are not yet fully integrated. The Enneagram itself is still an extremely esoteric (rare and obscure) field of study and self- discovery… So, where does this leave us?

Fortunately, with the wisdom and power in this book and other resources available on the web, you can still benefit from helpful teachings & perspectives of the Enneagram. We mentioned earlier that we would be looking into the psychoanalyst Carl Jung's *Universal Archetypes,* and how they relate to the intentions of the Enneagram. We are doing so below, but first it is important to bring into awareness the significance of our subconscious, unconscious and conscious self; as this helps provide an understanding into the nature of both self and personality. The diagram below may help you understand the various elements of your Enneagram Personality type better.

The Subconscious: The hidden influences which shape, create and influence your thoughts, feelings and choices. The subconscious mind is responsible for all aspects of your "self," psyche and personality.

⇩

The Unconscious Mind: The unconscious mind contains the things you are unaware of. It is linked to your subconscious, as the subconscious influences everything which goes into the unconscious and conscious minds. Repressions, self- denial and "shadow" aspects of the self and personality often manifest and transfer to your unconscious mind. (Although they begin in the subconscious.)

⇩

Preconsciousness/ the Preconscious Mind: This part of mind is responsible for the day- to- day memories which shape and structure the person you know as "self." (Yourself.) It is easily accessible and often only needs a question, brief attention or momentary shift in conscious awareness to access it. Thoughts and memories in the preconscious mind rest only a fraction away from your conscious mind and are not buried deep down, such as in your subconscious or unconscious mind.

***The Conscious Mind*:** This is conscious awareness, or full consciousness. It is the aspect of mind which is present, attentive, aware and in control- it is the part of your personality which makes choices and decisions. The conscious mind draws its information, awareness and wisdom from all of the "minds" before it. It is also strongly connected and tied to your subconscious- the part of you which is not fully aware but still influences your feelings and actions.

When looking back to the Holy Idea, Virtue, Passion and Ego-Fixation, this insight into the nature of mind and the psyche may help shed some light into your own desires, motivations and choices. A lot of what is taught in the Enneagram shares similar, if not identical, beliefs and perspectives regarding the nature of mind and self to many other traditions and schools of thought.

To conclude this book we will be looking at one of the most influential systems which can help us to understand our true selves and inner divinity. Although a psychologist and psychoanalyst, Carl Jung very much believed in the power of the human spirit, and that we all had a soul and "timeless essence" to us. This is made apparent through his advocations in his Universal Principles, which we will be exploring now. Understand that the point of learning this is to remember how the Enneagram is drawn from ancient religions, philosophies, cultures and traditions. Well, so are the Universal Archetypes!

Brief Background into Carl Jung

Carl Jung was a Swiss psychiatrist and one of the founders of modern psychology. He came up with a set of *universal archetypes* of *the Self*, which are aspects of the whole human being, relating to everyone on earth and everyone who ever will be. These archetypes are aspects of the *collective consciousness.* His archetypes were created from his explorations of the collective unconscious through different religions of the east and west, mythology and alchemy. He believed archetypes manifested themselves in dreams through symbols and figures, and these dreams could be used to understand daily life. Once activated, they could unlock a specific frequency or energy type within the person, specifically associated with the archetype. As our brains are transmitters and receivers of consciousness, in dreams we can learn a great deal about our impressions and beliefs and gain insight and wisdom for everyday reality.

The Persona

The persona is the image you present to the world in daily life. It is your mask or sense of image, such as "the caring empath," "the crazy scientist," "the gifted artist" or "the helpful healer." The term persona literally comes from the Latin *mask* and in dreams is represented by the *Self*, a character you know is you. Jung referred to the persona as "the conformity archetype," and believed it essential in personal development. Working with dream symbolism can assist self-discovery greatly, as through consciously connecting to your subconscious yourself and further asking it to show you some sign, message or insight into your persona; great healing and growth can arise. The persona is very similar to your Holy Idea and Virtue, or at least what they represent. For example, being a type 9 means that your higher self, or the best version of yourself (represented by the *Holy Idea*) is one of holy or unconditional love. As type 9 is also referred to as the *mediator* or *peacemaker*, and due to the Virtue of type 9 being conscious *action*, this signifies that your persona will be something in alignment with these.

The Shadow

The shadow is those aspects of self that you wish to deny. It is the hidden, repressed and rejected parts of your personality, character or being which you perceive as ugly, dark or undesirable. They can represent fear, anger or weakness and usually manifest as desires and memories which one simply does not wish to accept or integrate. Animal instincts and sexuality are integral parts of the collective shadow. In dreams, the shadow can represent itself as a lowly creature, such as a dwarf or an animal representing primal and instinctual urges, such as a tiger. The general rule is that the stronger we identify with our persona, the more we deny other aspects of ourselves.

Recurring themes that often accompany the shadow self:

A suppression of sex, love, and desire;

a suppression of intimacy, self-love, and connection;

a suppression of money, abundance, and financial prosperity;

a suppression of health, personal development, and self-mastery; and

a suppression of material bliss, new opportunities, and living one's dreams.

Looking at the shadow aspects of oneself can be a beautiful process as it allows you to delve deep into the positive associations of your character. Furthermore, the shadow can actually be dangerous when not recognized, as denying something leads to its greater persistence. We all have light and dark, good and evil, and opposing forces within us; therefore, by coming to terms with our own shadows and the collective shadow aspects you can begin to live life in a state of harmony, balance and contentment, truly connected to and at one with your unique gifts and strengths. The same or similar principle is true in the Enneagram; although they are called different things, the Shadow relates to your *Passion* and *Ego-Fixation.*

The Anima/Animus

These are female and male aspects of yourself, respectively. Everyone possesses both feminine and masculine attributes, therefore the anima and animus are not limited to being a woman or man. Empaths, for example, can naturally be seen as more feminine as empathy is to be able to connect with others on an emotional level, which is a feminine quality. Learning about your anima and animus is a key way to delve deeper into your personality and psyche. They are also doorways into understanding your Enneatype!

The Divine Child

The divine child is your truest self in its purest form in that it represents your innocence, your sense of vulnerability and your helplessness. It also, however, symbolizes your goals, dreams and aspirations, and your highest potential. In dreams, it is represented by a child, baby, or infant. Now, regarding self and spiritual development, we can see this universal archetype as a fundamental part of one's journey. If you are looking for an archetype to connect to and embody in your meditations, for example, the divine child is the perfect one. Why? Because empathy is literally all the things Carl Jung defined the divine child as. Innocence, vulnerability and helplessness (an empath's shadow traits), goals and dreams, aspirations and highest potential (and empath's purpose and true path): this is ultimately the journey of an empath. And whether you consider yourself an empath or not, *empathy* is one of the main goals of learning about the Enneagram for healing and spiritual evolution. The empath nature can thus be seen as the highest expression of self, as empathy us such an advanced and evolved frequency of emotional connection and expression; and the Divine Child archetype can help us get back in tune with our innocence and inner humility.

The Wise Old Man

This archetype is pretty self-explanatory. In dreams, he is represented by a masculine figure of some sort; a father, teacher or a masculine-type authority figure. The purpose of the wise old man is to offer insight, direction, and guidance. Through dream work, working consciously with our dreams, we can explore the subconscious realms and actively ask for wisdom, guidance, and assistance. Dreams can be seen as gateways to the soul and can be used for help in everyday life.

The Great Mother

Like the wise old man, the great mother is the feminine aspect of your subconscious who can be called on for help. She is the great nurturer and manifests in dreams as a grandmother, mother, fairy godmother, or any powerful and nurturing female figure. This archetype is very important when it comes to exploring our nature as the negative expression and association of the great mother is the *witch*. Now, we are not talking about good witches, those real medicine women or magical, crystal-loving fairy types, but the "evil" type. This type of witch is the terrifying, green goblin, stereotypical black-cloaked witch with an evil face. She represents dominance, death and seduction, and is very important in embracing the shadow elements of ourselves that we may wish to deny.

If the witch shows up in a dream, it is a signal that we are *rejecting or suppressing* the aspect she relates to. Themes of *death and destruction, seduction and sexuality are an integral part of life*, even if not acting upon them (sex, desire, lust, etc.). So, like with the shadow, ignoring a part of our nature in pursuit of perfection or to be everything for everyone creates an accumulation of suppressed energy to occur (which manifests itself as the "evil witch"). Nightmares show us just how important it is to accept and integrate those aspects of the self we do not necessarily wish to embody, as they are our subconscious going on a sort of "attack." (Or feeling like it is being attacked.) It is only when we truly do this- integrate and accept- that we can become the best versions of ourselves and fully embody our strengths.

The Trickster

As the name suggests, the trickster is the practical joker of the collective consciousness archetypes. He may present himself in dreams if you have taken yourself too seriously, misjudged or over-reacted to a situation or person, and can once again provide insight into if or when we have been falling prey to our shadow and unhealthy traits.

The Ego

The ego is one of the main archetypes of the personality and is the center of consciousness. It is the "I," the aspect of self that is central in daily life. It is the part that relates to the psyche and all things personal, personal experiences, lessons and learnings throughout the entire human journey. The other part of the psyche is the unconscious and through the ego we learn about our inner workings and currents.

The Self

The self can be seen as the most important archetype in that is the *whole personality*, the totality of it. It is centeredness, the union of the conscious and unconscious and embodies the balance and harmony of the opposing elements of the psyche. In dreams, the self is depicted as a circle, mandala, crystal or stone (impersonally), or as a royal couple, divine child, or some other symbol of divinity such as Christ, Buddha, or other great spiritual teachers (personally). These symbols of the self are all representations of *wholeness, completeness, unification, and reconciliation of opposites*, therefore, can be connected to when discovering the deepest workings of your psyche and self.

The self, your whole and integrated personality which can be seen as *self-mastered*, is your inner guide to help navigate you through all of life's queries, concerns and problems. It is essentially your superpower! By taking the necessary steps and following all the various insights and pieces of knowledge throughout this book, and the profound meditation exercise in the *Bonus Chapter*, you can live your best life full of joy, content and abundance, and achieve your highest possible vibration; further maintaining it. The Enneagram offers a sense of completion and "*coming full circle*," coming full circle within yourself and for yourself.

BONUS CHAPTER

A Unique Meditation & Self- Healing Exercise

Now you are familiar with your Personality type, the subsequent Holy Idea, Virtue, Passion and Ego-Fixation, and how the Enneagram itself can be used to develop empathy, meditation and mindfulness; it is highly significant to *actively engage* in self- development in relation to the teachings and integrations of this unique system of self- discovery. Below is a simple yet powerful meditation and self- healing exercise, allowing you to connect to any of your positive and favorable personality aspects, and Holy Idea or Virtue simultaneously, to increase, enhance and embody each for intended effect. Knowledge is power, but the active application of the knowledge and wisdom contained in the Enneagram is what will really amplify and enable you to access it's teachings.

Step 1: Creating A Sacred Space

The whole principle of the Enneagram is based on the concept of there being a divine and timeless reality. This can be confirmed through the frequent advocation of each of us being made of "Essence" and also having a soul; the soul is timeless and infinite and further transcends any perceived limitations of the physical realm and material reality we reside in. "Sacredness" is honoring and recognizing the sacred aspect of life, that there is a divine and spiritual reality awaiting us. But, this reality is not separate from us or something seldom reached, it is very close to us and easily accessible with the right willpower, openness of mind, heart and spirit, and willingness to surrender and let go; detaching from the physical body and opening our minds to worlds of limitless possibility.

A sacred space, therefore, is essential for a successful and profound meditation.

Start creating your sacred space by setting up a physically comforting and safe- feeling space. This may include cushions or soft pillows on the floor, and "clearing" your room or space performing the meditation in. Clearing is also known as energetic clearing, or energetic cleansing. It is about recognizing the subtle- spiritual energy which flows through all living things, including our own bodies, and all physical environments. Your environment needs to be clean, pure and cleansed in order to experience this meditation exercise in its optimum. So clear and cleanse the room or physical environment, and *set some intentions* in the space, thus creating an element of *sacredness.*

Focusing on the elements, the elemental energy permeating around you is the next step. Earth, air, fire, water and ether- or spirit- construct physical reality as we define and perceive it. These elements flow through our own veins just as they flow through planet earth and the universe as a whole. Elemental energy is essentially in everything, from the fruits and soul- nourishing foods we eat, the water we drink and bathe in and natural world we walk through. *Earth* can be connected to do by having a crystal, special gemstone, shell or stone near you. *Air* can be present with a feather. *Fire* can be near you with a candle, and *water* can be attuned to with objects from the sea or a simple glass of water. For *spirit or ether*, any symbol which reminds you of the divine or any deity you pray to will suffice. Alternatively you may have a picture of some spiritual or healing symbolism, such a the 7 major chakras or some other ancient symbolism. Whatever helps you connect to the innermost core of yourself, to your essence; use.

Get comfortable and set some intentions. Intention setting is very important for this meditation to work, as through the power of intentions you project subtle energetic influences through your thoughts and essence. Intentions come from your truest self, your higher self and your heart and soul. When you set intentions from an aligned and sacred space you realize that life is a unified and interconnected experience, and that you can manifest things in harmony with your heart's desires when thinking, perceiving and acting from such a pure space. Our hearts give off powerful yet subtle vibrations, and the frequencies of our thoughts emanate these subtle vibrations. In short, we are powerful shapers and creators of our realities- both our inner worlds and outer external ones.

Next, make sure that the lighting in the room is such that doesn't distract you. Too much or bright artificial lighting can be detrimental and interfere with your ability to connect to your intuitive, higher self essence. Light some candles, safely, or opt for a natural source of light such as a Himalayan Salt Rock lamp, or an eco- friendly LED Moon lamp. Alternatively, dim the lights or just have a small lamp nearby.

Finally, physically "cleanse your space." Organic and herbal incense, sage, frankincense resin oil, or an essential oil in an oil burner are all highly effective ways to cleanse your space. Sage, frankincense and a Peruvian wood called Palo Santo are particularly revered by many spiritual practicers and frequent meditators.

Step 2: Define Your Intentions: What is the Purpose?

At this stage it is important to become aware of the purpose and what will actually be happening. This was briefly shared in the introduction, however let's re-clarify and expand. This meditation allows you to connect to any of your positive and favorable personality aspects, and the Holy Idea or Virtue simultaneously. So, during the meditation there will be a technique of visualizing your intended quality or positive personality attribute and watching it grow or expand within you. The technique of visualization is very powerful, as on a subtle level intentions, frequencies of thought and belief, and the essence and meaning you give something actually hold considerable power. There is a reason why the saying "dream something into being" has become so popular. It is because our minds are powerful tools and channels for manifestation.

Whatever you wish to amplify and integrate can be achieved through recognizing the influence and power of subtle energy and intentions. Your thoughts hold great impact and can be catalytic to your self- development and spiritual growth.

Step 3: Familiarize Yourself with Your Unique Holy Idea, Virtue, Passion and Ego- Fixation.

Before you officially sit down in your sacred space and get ready to go deep into the meditation, it is very important to make sure you are completely comfortable and familiar with the essence of your Personality type. Read, study, learn and memorize. Why? Well, if you don't know what you are connecting to, or why, you won't be able to actually do it. Again, knowledge is power and you need to make sure you are in the know. Know your strengths, weaknesses, gifts and talents, and shadow traits, for this is the only way you can use this exercise for healing, growth and personal transformation.

The Meditation

You may wish to record this on a mobile or recording device. Speak in a soft, compassionate and soothing voice, or just speak as if you are empathically giving yourself direction and guidance. This meditation can last from anywhere between 15 to 45 minutes, depending on how deep you wish to go and how long you feel comfortable being in such a transcendental space. Or you can just familiarize yourself with the meditation and engage in it without spoken guidance.

Start by closing your eyes and going within. Calm your mind, still your senses, and feel safe, secure and protected in your body. You are protected and in tune with the world around and your inner being. There is only security, guidance and protection here, so begin to breathe and feel relaxed inside.

Any tension or resistance that may appear is natural. The key to know with meditation is that a lot of past memories, pains, trauma and wounds can be (unconsciously) stored in our cells. This means that we begin meditation, if we are new to it, the process of conscious breathing itself and filling ourselves up with empty space can trigger natural tension. This is perfectly normal, so instead of resisting, panicking or feeling anxious or fearful, just focus on your breath and surrender. Any scary, negative or unwanted thoughts will flow through you just as your breath is flowing through you.

Now you should feel more settled and content within, keep breathing consciously. With each inhale visualize a beautiful, gentle but warm and powerful golden light. This light is entering through your first center- your stomach region- and travelling up your spine, past and over your heart, and into your head. As it makes its way into your head, either in between your eyes above your brow or directly at the top of your head, picture this same golden gentle light making its way out of you. Exhale and see this healing light leave you. As it leaves, know that all tension, worries and concerns are leaving you- you are being cleansed with this golden light.

Continue this process for 5- 10 minutes, keeping your focus on your inhales with the light energy travelling up through you, from your gut, and spilling out into your head; to then be released through your exhale and along with it any tension and negative thoughts. At this stage, be mindful of your stomach, your gut and your center and its significance and symbolism. See this beautiful healing light filling you up inside, shining light and health on your Passion and Ego-Fixation, and any undesirable traits which may come with your Personality type. Like a DNA spiral, or a snake, visualize this same light with each breath making its way to the crown of your head, filling up the forefront of your conscious mind- your brow- and leaving you. Repeat this as many times as feels comfortable, and until you feel light and content within.

Now you are going to *fill yourself up* with the quality you wish to expand and embody. This can also be called *energizing-* you are essentially energizing yourself with your positive and favored traits and characteristics. This element of the meditation can be done as many times as you want and feel comfortable doing, for example you may choose to start with your Holy Idea, then work your way onto your Virtue, and then finish with 2- 3 different personality strengths. For example, if you are Personality Type 9 you would visualize and project "love" (Holy Idea), "positive or empathic action" (Virtue), peace, harmony, empathy or compassionate and mindful communication (some of your qualities). *Empathy* can also be used for *all* 9 Enneatypes, and due to the purpose and intentions of this book, we will be sticking with empathy as a quality in the remaining guidance.

Start by taking some deep breaths and reconnecting to your inner core once more. Set a brief yet strong intention of connecting to the quality or trait you wish to embody and amplify. See the word, connect to the essence of the quality, and mentally project a visual representation of it. Now, picture the word "empathy" (holistic example) in your mind's eye whilst breathing in from your gut. This time, visualize a beautiful green light growing and expanding from your first center- your gut- and travelling up your spine, through your heart and out into your Third Eye Chakra (energy center). As described earlier, this is the seat of your intuition, psychic or spiritual vision, and connection to the divine. It is in this energetic center where your consciousness can become activated in new and profound ways, and your beautiful qualities can be enhanced through visualization and mental imagery projection exercises. Continue to see this green light, linking to the heart and quality of empathy, energizing you from the inside. As you exhale, visualize the same green energy light leaving you and dispersing into the ether and spiritual realm around, with a subtle intention of releasing all that no longer serves you. You don't have to actively picture fears, worries or negative thoughts leaving you, instead just picture the healing green light whilst setting an intention for healing and release. This in itself will assist in removing any thought, feeling, emotion or belief not alignment with your highest self, whilst keeping the focus on the positive characteristic or personality trait you do want to energize.

Once a steady flow has been created and you feel connected to the green light, begin to develop a deeper connection the quality wishing to be increased and integrated. Start to envision scenes or scenarios in your mind's eye with the quality in focus. They don't have to be too complex, just make sure you connect to the feeling and essence of the visual representation. For example, for empathy you may picture yourself smiling with a warm glow around you whilst you offer loving and supportive words to a stranger. You may envision you actively engaged in a charitable or deeply helpful service or action, or you may see yourself walking through nature smiling at the birds, flowers and trees, in deep recognition of the conscious life force flowing through all living things. Any scene which resonates with your core and true self, attune to and align with.

Stay in this energetic space connecting to the feeling for a few moments. Continue to breathe consciously, and with each inhale visualize this green, warming and comforting empathic light filling up the scene in your mind's eye.

Now you are going to bring your hands up to your face, a few inches extending outwards, with your palms facing you. In many ancient and healing systems the palms are known as the palm chakras, and it is here where many healers and practitioners of healing energies channel energy and universal or divine power. Set a brief intention, and run your hands all the way down from your Third Eye/ brow to your gut/ stomach. The intention is to *cleanse* and *clear* yourself, bringing you back to point or ground zero. The purpose of this is due to the next stage in this self- healing meditation.

Just like with "filling yourself up," this time you are going to *clear yourself* and heal your soul and psyche of your Passion and Ego- Fixation. Please note this is not about denying or repressing your Passion and Ego- Fixation, the intention is solely to transcend the limiting and self- destructive beliefs and energetic associations connected to it. For example, as a Personality Type 9 you will be focusing on Slothfulness and Laziness, and how you wish to overcome these negative character traits. So, to begin, bring your left hand a few inches away over your gut. Hold it there gently and move it back and forward slightly, until you feel an energetic pull like a magnet. You should be able to feel this subtle magnetism quite powerfully at this stage in the meditation. Gently rest your hand, with your palm facing towards your stomach, a few inches away and bring your right hand up to your right shoulder. Keep your palm facing outwards and make sure the location of your hand is about shoulder's length, just above your heart. Take some deep breaths and start to feel a gentle force like a circuit between the palms of your two hands.

Now, project and see your Passion shimmering from your gut and stomach, and slowly emanating out into your left hand's palm. Take some deep conscious breaths and with each inhale, visualize this Passion leaving your gut and spilling out into your left palm, whilst keeping your mental focus on the intention of being free of it. As you continue to breathe in, see this unwanted Passion circling round all the way to your right hand and with each exhale leaving you. Watch this Passion flow through your right palm, spilling out into the ether where it will be healed, cleansed and released. Your Passion characteristic is being cleared and cleansed, leaving you with each breath. It is returning back to the divine where the divine's unconditional love and compassion is sending forgiveness, healing and acceptance.

Continue to do this until you feel light, airy and with a content emptiness inside. You may be feeling an energy current or strong sensation now, or you may physically see swirls of light surrounding you or emanating from your stomach and hands. This is all completely normal, so just be still inside and know that you are safe, blessed and protected. In this energetic space all is well and you are loved, and cherished. Great and deep healing and transformation are occuring now.

Repeat this with the Ego- Fixation and anything else in relation to your Personality type you wish to let go of.

Finally, and to end this powerful exercise, become still inside once more, connect to your physical body, and bring your awareness to your physical surroundings. Keep your eyes closed still but be mindful of the sacred space you have previously setup and your physical environment. Become conscious of any physical sensations which may be arising, within or around you in your subtle body, and once again create some intentions. Thank the power of the Enneagram and the power of the universal energies which assisted you throughout, and finally bring your awareness to the sequence of visualizations and techniques throughout the meditation.

When you are ready, open your eyes and know that you have just been blessed and filled with your Holy Idea, Virtue and beautiful Personality traits, and simultaneously cleansed and healed of your Shadow traits. All which no longer serves you are leaving you now and healing and transformation are still occurring on a deeper level.

BOOK 2
THE SACRED ENNEAGRAM

A Journey to discover your unique
path for Spiritual Growth and Healthy Relationships

Andy Connor

CHAPTER 1

The Enneagram

Enneagram Origins-in brief

The precise origins of the Enneagram are a disputed mystery. It is possible that they go as far back as the ancient Greeks.

The word Enneagram derives from the Greek word ἐννέα *ennéa*, which means "nine" and the Greek word γράμμα *grámma*, which means something "written" or "drawn". In fact, the nine-pointed Enneagram symbol, today, is used to represent the nine basic personality types, known as "enneatypes" and enable a better understanding of ourselves.

The enneagram is a diagram, which is believed to have first come about in 1915 by a philosopher teacher George Gurdjieff who used it to aid his students in Human Development courses. Gurdjieff claimed then that the enneagram was a secret ancient diagram. Ouspensky, one of his Gurdieffs students recalls Gurdjieff using the enneagram to represent "the fundamental laws of transformation and the organic unity of everything existing". Later in the 1960's, Oscar Ichazo, added 9 personality types to the diagram and later still, Claudio Naranjo, combined it to modern psychology. The personality types already existed, but were then unified into one unique system.

What Is the Enneagram?

The Enneagram is a nine-point diagram and every point represents a personality type. The diagram consists of three elements; the external part in the form of a circle, the internal part is made up of a triangle and an irregular hexagon, which are inside the circle.

Most people in todays' modern the Enneagram as a "tool" to gain success in the business environment or to learn how to get the most out people. The enneagram, however is not a simple personality test, it is on the other hand, a system, which is studied to enable spiritual depth! The Enneagram alone is only a part of the puzzle, a stepping-stone to aid personal awareness or even spiritual awakening.

Self awareness and our 3 centres

Do we really know who we are? I'm sure most people, at some point in their lives, have asked themselves "who am I?" well, the enneagram is a part of a process we can use or study to help us understand just that.

Think about when you have been in a classroom or a meeting and have been asked to introduce yourself and say something about yourself to the rest of the group. We usually all begin with our name followed by what we are studying or what our job is. Does this mean that we believe that our job defines who we are? How we make a living is a tiny part of our identity, and sometimes doesn't come near to reflecting who we really are. We become so used to giving this information about where we have been or what we do for a living that even we start to believe that that's who we are. If we let our academic achievements or failures define who we are, then we will not recognize out true self and before we know it, our achievements or failures will claim our whole identity.

Our whole identity however, is a lot more complex, we are not what we do or what others think of us! We are also not what we have. People define themselves as happy if they have a successful job, a wife, children, a beautiful home etc. how do we disconnect ourselves from out false identities? This is where the enneagram could help us rediscover our identity or essence.

We are made up of three centres of

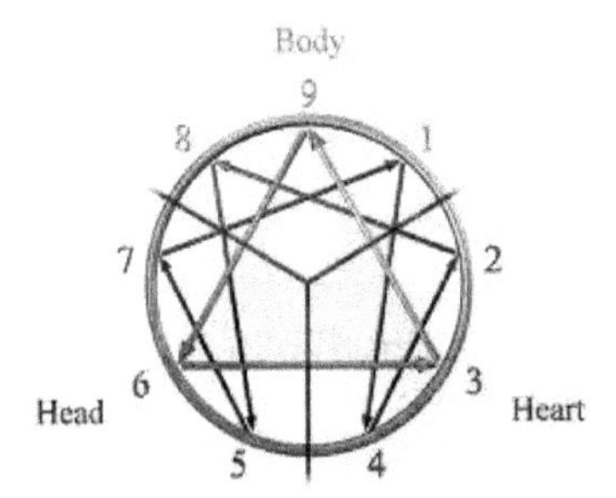

intelligence; the head, the heart and the body/the gut. In fact the nine personality types are split into three triads:

- Gut (Instincts): Types 8, 9, 1
- Heart (Feeling): Types 2, 3, 4
- Head (Thinking): Types 5, 6, 7

Fig.1

The 3 centres of intelligence are represented in the inner circle of the Enneagram above, divided into three triads.

Over time our centre becomes lost and our thinking, feeling and doing (heart, head and body) become unbalanced by nine principle traits/personality types and we forget or distance ourselves from the connection we have with our essence. These nine personality types are represented in the enneagram, and all nine types contain the three centres of intelligence, however one will prevail over the others.

By understanding our personality types through the enneagram, we can try to get back to our essence by escaping the behavioural loops we get into which are due to these personality types/ego.

Personality vs Essence

We are all made up of "our inner self" or essence and our personality, although we are born with our essence, our personality is acquired or learnt. For this reason, we commonly hear the concept of the word Essence being compared to a child. As children are less exposed to external inputs or teaching, they are more connected to their inner essence/true being rather than their personality, also referred to as the Ego. "A small child has no personality as yet. He is what he really is. He is Essence. His desires, tastes, likes, dislikes, express his being such as it is." (*in search of the miraculous fragments of an unknown teaching P. D. Ouspensky)* The essence has no need for words, words are associated with a persons' personality and small children do not use language to express themselves. Babies are born being simply their essence; their personality is created and grows by external influences and by imitating the people around them. When education begins, so does personality as we are taught how we are "supposed to be".

A persons' personality does not light up at the sight of a cat playing in the street or at watching a leaf fall off a tree during Autumn, the essence of a person, on the other hand, does. We can learn a lot from observing a small child as they glare with joy and excitement as a leaf magically floats to the ground or witness their whole face light up as they watch this beautiful, colourful creature (cat) rolling around on the grass. The more modern our culture, the more of our essence we lose, we become distracted and unable to be present. (*"We have to distinguish between a man as he is in essence, and as he is in ego or personality. In essence, every person is perfect, fearless, and in a loving unity with the entire cosmos; there is no conflict within the person between head, heart, and stomach or between the person and others. Then something happens: the ego begins to develop, karma accumulates, there is a transition from objectivity to subjectivity; man falls from essence into personality."* (Interviews with Ichazo, *page 9)*

Our external ego is useful when socializing and communicating with others. Having said that, we are unbalanced when our ego overshadows our essence. Our actions and behaviour become automatic and rigid actions and reactions leaving little space for our essence or whom we really are. The Enneagram teaches us about the dynamics of our personality traits in order to help us with our journey to become ourselves again and escape the loop of our behavioural patterns based on our personality traits.

CHAPTER 2

Centres and Personality Types

Intro - Centres of Intelligence

When we hear people referring to Intelligence, they are often referring to intellect, deriving from one's mind, however, the heart, the head and the body are all sources of intelligence. In fact, the three centres of intelligence, which need to be balanced, are:

The heart centre- this is the emotional centre, used to connect to our positive and negative feelings and to understand that of others

The body centre- this is the instinctive centre, the "doing centre and is situated in the gut

The Head centre- this is the rational centre

We all possess one dominant centre, if we don't open up to all three centres and become stuck with one centre dominating we will miss out on our whole self and are unbalanced.

The Nine personality types

People of the same personality types view the world in a similar way and interact with others similarly. The reactions of people from the same type may vary depending on parent's types and cultural values and background.

The nine personality types that are represented by the numbers 1-9 (as you can see on the previous page) on the 9 pointed Enneagram are; the protector, the peacemaker, the perfectionist, the observer, the sceptic, the enthusiast, the giver, the achiever and the individualist. Despite identifying with one enneagram "type" more than the others, many people could find elements of themselves in all of the types. The type that matches how you are the best is known as your basic personality type.

The personality types are defined by a set of dominant behaviours, motivations, and fears. The systems' aim is to allow an in-depth understanding to your type. By analysing the strengths and weaknesses of the type and becoming more self-aware, the process of self-improvement and reaching your full potential and spiritual awareness may begin.The nine personality types are represented in the enneagram, and all nine

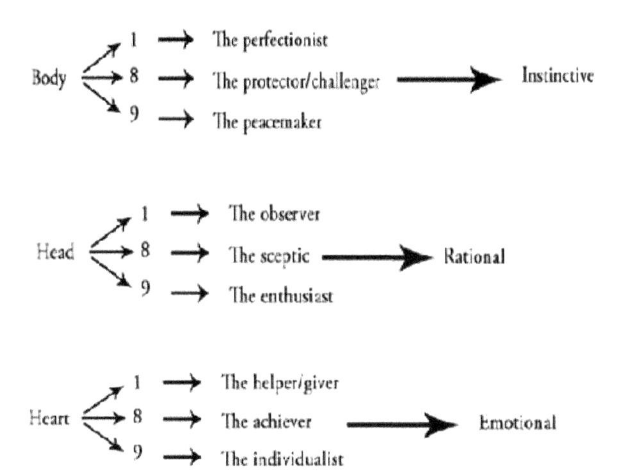

types contain the three centres of intelligence, however one will prevail over the others.

Each centre of intelligence contains 3 personality types (see fig.1)

Finding your Enneagram type

The first step to understanding which Enneagram type we are is to start from your centre. Knowing your centre before looking for your number simplifies things. So, you need to pay attention to how you operate, is it from the gut, the heart or the body? (Looking back at the diagram on the previous page will help) Are you more instinctive, rational or emotional?

The Body Centre

People who possess a predominant body centre will react more on gut instincts and physical sensations, they are full of energy and very much alive. They tend to act instinctively without thinking things through first, and usually relate to others in a direct and straightforward manor, which also makes them confrontational towards others. The body centre is correlated to anger, when people of this centre express their anger in different ways, others can sometimes view this as being aggressive. This type of person will mostly reflect on their actions after having reacted instead of reflecting before acting.

The Heart Centre

People who respond to feelings, emotions have a more powerful heart centre. Heart centred people pay attention to their image and are relationship orientated. They tend to seek recognition and approval from others, other peoples' reactions and responses towards them seem to have a direct consequence on their emotions. They are also able to adjust their image depending on the situation or environment around them.

The Head Centre

People who rely on logical thinking to situations are head centred. They are rational and prefer to think thinks through thoroughly before acting. They don't usually throw themselves into relationships with others and prefer to observe at a distance before interacting, as they like to know what to expect of a situation before they have to deal with it.

Once you have understood which your strongest centre is, you then need to look at the 3 types in that centre to identify which type suits you the most.

The three principle ways of interacting with the world are included in the Enneagram system, they are: feeling, thinking, or doing. The nine numbers are divided into those three categories (referred to as triads). The triad that most reflects how you operate is determined by the way in which you perceive information or situations.

So, if you are Heart centred you will need to look at types two, three and four. If you are Head centred you should examine types five, six and seven, and if you are body centred you should pay more attention to types eight, nine and one.

Identifying your personality type

When trying to identify ourselves with a personality type by matching up the way we behave and how we see ourselves with these personality traits, determining which one suits us the most can result confusing. The biggest obstacle when trying to do this is that most of us will be able to see traits of more than one personality type in ourselves. Maybe we are convinced we are type 1, but then go on to read a description of a type 9, and identify ourselves with one or two aspects of that type. It's very easy to become confused at this stage, and some people use the wrong category to define themselves. This happens in general when trying to fit in with a stereotype. Let's not forget that The enneagram, however, is not a simple personality trait test, but a system which we could use to help our growth and self-awareness. When considering our personality, we have a lot to learn from the enneagram, the way we act in our daily lives and how we interact with others are merely consequences of our motives.

When we talk about motives with the enneagram, we really mean our common fears. Each enneagram type (from 1 to 9) shares a basic fear. It is what we most fear in our lives that drives and motivates us to act in certain ways. For example, a type two, is always acting to help others, they do this because their basic fear is that of being unloved or unwanted. Therefore, if you can truthfully pinpoint your main basic fear, which you can see in the 9 types description to follow, then you will most probably reach your type with a lot less confusion. You may even have more than one fear, but it's easier to pinpoint your main fear than one single trait. Your one basic fear will be the one that scares you the most, the one that deep down, maybe unconsciously you try to avoid at all costs.

CHAPTER 3

Personality Types For The Body Centre

Type One- the perfectionist Overview

This personality type focuses on doing things "right", is a critic of both themselves and others by setting high standards, and strive for self-control. Type ones are honest, reliable and responsible people who enjoy correcting anything wrong. They tend to categorize everything into black or white/wrong or right, with a constant need of self-improvement and see fun as disorder.

Their negative characteristic is their common emotion of anger. When Ones experience anger, they seldom express it, they try their best to keep their emotions under control. They frequently become irritated and feel resentment towards others who they see as irresponsible or not working to a high quality standard. Although ones are hard on themselves and extremely self-critical, they do not take well to others criticizing them or pointing out a mistake they have made.

Basic fear for a one type- Type ones act in this way because their basic common fear is being evil or corrupt.

Self-improvement for a Type One

People with type one personality are known to be quite serious and straightforward. Their high internal standards and conscientiousness can come at the cost of their personal needs and desires. If you are a type one, aka a perfectionist, there are a variety of things that you can do to develop yourself and attain growth in different areas of your life.

Learning to relax

Getting a perfectionist to relax is akin to getting blood from a stone. They are hard workers and will rarely take time out for themselves. Regardless, a human spirit can only do so much before it is spent. That is why self-nurturing is of utmost importance for people with type one personality if they are to keep things on an even keel.

For starters, ensure that you take some time out for yourself every day to engage in some recreational activity. Since planning comes naturally to you, mark it a daily task to enjoy something other than work. This could include playing your favourite sport or watching that movie that your friend has been meaning to watch with you for a while.

Fortunately, the world will not enter the dark ages if you do what you want to do, even though it may sometimes feel as if it might. So, learn to listen to what your mind and body want from you and do it, even if it is on a whim. Slowly but surely you will find out that there is a whole world inside of you that has been neglected and is bursting to come out.

Appreciating the funny side of life can also lead you away from your worries, even though for a moment. Memorize a few jokes and crack them in front of your friends. This will allow you to wean away from your black and white worldview and induce some colour into it.

Sometimes when life becomes too much to handle, take a step back and simply observe. Meditation comes in especially handy at such instances and works like a charm to reduce stress. Take a yoga class to manage your physical discomforts and let your body go through the motions. Soon you will realize how important your body is as a vessel and your mind will thank you for it.

Do not be afraid of taking a vacation after a few months of hard work. You may think that it may not do much but being physically away from your work and your worries can allow you to re-establish connection with your inner self.

Understanding Anger

Anger is a common response when we feel we've been slighted or when the shortcomings of others bear hard upon our own sense of morality. Though you may feel that you have a lot to teach to other people, do not expect them to change overnight. Things that may seem obvious to you may not be so obvious to other people, especially if they are not as conscientious and self-motivated as you are.

Everyone wants to do what seems right to them and though they may agree with you on principle, they may not be inclined to do what you want them to. In such situations, it is not wrong to get angry. But understand that your behavior might prove to be a better example than your words could ever be. So, have patience when dealing with people.

Many times you may find yourself injecting your cynicism and sarcastic comments when talking to people. More often than not, it is a sign that you feel hurt or are being defensive. Frustration at being unable to live up to your expectations will only fuel your anger and though it is a natural human emotion, when it goes unchecked, it can lead to pathological behaviour patterns that may end up jeopardizing your relationship with other people.

Try to realize that changing one's perspective is easier said than done. So, whenever you feel that you are harbouring feelings of resentment or anger, learn to channel them in some positive way. Talk to a friend or start a journal and let go of all your negative emotions that have been poisoning your mind. At the same time, make it a point to forgive other people who you feel have caused you harm. It is one of the easiest ways by which you can lift yourself out of your own toxic pattern of behaviour.

Maintaining a work-life balance

Being a perfectionist can be both your greatest gift and your biggest flaw. That is why you need to understand to what extent you should pursue excellence and when should you let go. The type of job you have will greatly determine this, so it is crucial that you assess if your work is fulfilling enough for you.

At the same time, don't overload yourself with work. Do only what is designated to you and don't take on others' work fearing that they won't be able to do it as well as you. To ensure that you don't work yourself into a rut, allow yourself to do a few things as fast as you can no matter how imperfect it may seem to you. You have many other obligations such as your family and your own well-being, which shouldn't be overlooked.

Ultimately, you want to understand the things that are within your control to change, to accept the things that you aren't and better still to know the difference between the two. Whenever you find yourself brooding over past mistakes, take a step back and instead remember the things that you did right and are most proud of.

As a perfectionist, it can be hard to overlook one's mistakes and imperfections. But, if you follow this chain of thought down to its logical extreme, you may end up feeling incapacitated and without hope. Instead, mediate your judgment between fairness and forgiveness and share your responsibilities with others. Most importantly, have fun in whatever it is that you are trying to do.

Relationship Goals

There is beauty in imperfection as much in perfection, and you should give yourself the chance to explore both sides of the coin. In your relationship, try not to be too demanding of your partner all the time. Give him/her some leeway when and where it is due. Do not overly criticize them for their habits, for they are simply being who they are, just as you are.

Every personality is different and has something unique to offer. Therefore, it is incumbent upon you to accept people for who they are. Fights and quarrels over petty things are inevitable in a relationship but if you can accept them with compassion and learn to relax with them, your time together will open up new vistas and riches of life that can be experienced fully.

Instead of flaunting your own achievements and putting your partner down, cultivate their strengths. At the same time, do not get stuck in doing what has worked hitherto though it may feel like safe territory. Instead, do things that you wouldn't do otherwise and bring some fun to your relationship. Novelty is the surest way to ensure that your relationship stands the test of time.

Though you may feel an irresistible urge to correct your partner at every turn, refrain from doing so. Know that rightness and correctness are not part of the natural order. If it were, we would all turn out to be more or less the same. Switch off your internal critic when you are with your partner and simply learn to unwind in their presence. Have compassion for yourself as well, for that will foster empathy and make you see your partner in a favourable light.

Always be grateful for your partner. It will help you to refocus your energy and will also broaden your focus of attention. This will also allow you to be grateful for little things in life and bring your mind from brooding over incessant worries back to the present moment.

On Parenting

Everyone wants their children to be the best that they can be, but type one personality can be a bit too idealistic in this sense. If you have children, you may want them to do things a certain way – put more precisely, your way. This can include observing colour schemes, not allowing toys outside the playroom, or strictly debarring food in the car. Though a few rules and restrictions are required in parenting, you should ensure that you don't overdo things to the point that your children become your slaves.

Self-esteem can only be cultivated in children if you allow them to do certain things their own way. It is alright if you want to be a proxy for society and establish a framework in which they can work. But, refrain at all costs from spoon-feeding them or constantly criticizing their shortcomings. This can do irreparable damage to children recovering from which may take their whole lives.

Children can be disorderly and messy but a little spontaneity can go overlooked. You may have to fight the urge to form your judgments and cut them some slack, especially if your children have been responsible, even by your standards.

As a perfectionist, it is in your nature to get things done correctly and with purpose and it can be quite hard to change your perspective towards things. But the personality of a child is still in its formative years and isn't as hardened as yours. Therefore, do not expect them to do everything with a set goal or a purpose or even to do it 'the right way'. Children learn by trial and error and all you can do is to provide a safe environment in which they can interact and learn on their own.

Finally, love your children not just for the things that they do right, but also for who they are. You can guide them and teach them things that you have learned, but you should abstain from shoving your perspective down their throats. Not only will this make your children love you more, but it will also allow you to have respect for their own world view.

Doing things that you normally wouldn't do

Sometimes the best way to ensure that you don't revert to your negative behavioural patterns is to mix things up and do things that you normally wouldn't do. For instance, report to work late one day and see how you feel about it. You may end up cursing yourself for taking such a step. But it is equally likely that other people will see you as a human who is as prone to the changes in circumstances of life as they are, and become friendlier to you.

If you usually reserve a table before dining out, try to leave things to chance and see what happens. Perhaps you won't get a table, and perhaps you may find yourself trying out a different diner and expose yourself to a different experience. Let the chips fall where they may so you can learn to work around situations that do not seem favourable at first.

Doing things out of the blue and without a proper plan is one way to ensure that your life doesn't become too predictable and boring. Leave the dishes in the sink, or don't take out the trash if that's what you are habituated to do. Or laugh off criticisms to see how people react. This will let them know that you are not as transparent as they might have come to believe and improve your stature in their eyes.

Make an impromptu plan with your friends and take a trip out of town just to see what awaits you outside of your comfort zones. More often than not people get frustrated with life because there is nothing new happening in their life and console themselves by thinking that it is good to be safe, to have a routine, even though deep down they may be urging to throw caution to the wind, even for a moment.

Make yourself available to people whom you normally avoid. This will show them that you too can go out of your way to appreciate what they have to offer, instead of being carefully wrapped up in your own bubble. You will realize how important it can be to meet different types of people and learn from them.

Positive things to say to yourself

Whenever you feel the negativity in your life is getting too much to handle, remind yourself that this is not the end and that you are much more than a simple conglomeration of neurons and behavioural patterns. Here is a list of things that you should say to yourself to get away from your negative mental state:

- It's OK to relax and not do anything.
- I am perfect the way I am.
- I shouldn't take myself too seriously.
- I will make my needs known to other people.
- I can find beauty in imperfection as well.

- I should be less critical of other people.
- It's OK to be less than ideal.
- Most things in life aren't black and white.

***

Type Eight- The Challenger

Overview

This personality type is loyal, energetic and protective. They feel at ease when they are in control of a situation, it makes them feel safe because they believe that they won't be left venerable or be attacked if they are seen to be in a position of power or in control. This is also, why they choose very carefully, when and to whom they show their most tender emotions, they don't like to be exposed or show their weaknesses.

They have an "all or nothing" approach to life and when they have a new project, all of their energy is directed there. Eights are bighearted and protect the they care about. Being body centred, they trust their gut instincts and are able to make decisions quickly. Although eights do not like to see weakness and incompetence in others, they do however defend people when they believe they have been unjustly criticized or attacked. This personality type does become angry quite quickly, even though they are quick to move on.

At their worst, they can be seen as domineering and aggressive, as they are confident and direct. Therefore, others can see them as intimidating. Eights love being in control and do not respond well if they are obliged to do something, whereas they expect others to follow their lead and to conform to their ways.

Basic fear for a eight type- Type eights act in this way because their basic common fear is being harmed or controlled.

Self-improvement for a type eight overview

Type eights are assertive, strong, and self-confident. They are also quite decisive, straight-talking, resourceful, and protective. However, these qualities also possibly lead them to be domineering and ego-centric. At times, type eights might feel the need to control the environment, especially the people around them by becoming intimidating and confrontational. They might also have issues with their temper, which makes them quite vulnerable.

Eights are known to possess incredible vitality and willpower and they feel most alive when these capacities are exercised in the world. They make use of their abundant energy to effect changes in their surroundings or leave behind a mark, especially on people around them. At an early age, they understand that it will require endurance, persistence, will, and strength – the same qualities that they would develop and expect others to have as well.

On their best, type eights make use of their strength to improve the lives of others, thereby becoming inspiring, magnanimous, and heroic. If you're an eight and you wish to come back to this, you can follow some of the tips mentioned here.

Relationships

Mostly eights are sought after as partners because they are strong, capable, and confident. People around them are reassured by their solidity and feel that they will offer them stability and protection in the relationship. they also display a lot of charisma – the instinctual energy is tremendous and people feel attracted to their intensity. However, one aspect that people tend to stay aware of is the fact that type eight might assert their energies a bit more forcefully for liking. Most people often create problems in their relationships.

If you are a type eight, you don't want anyone to control or have power over you, whether the power is financial, social, sexual, or psychological. Much of this behaviour reflects how much power you can retain for as long as possible.

You tend to become uninterested in other's feelings and self-absorbed due to feeling overwhelmed by your own feelings. Moreover, you tend to overreact to perceived rejection by losing your temper or withdrawing.

You push others for a more genuine response and become emotionally and remotely unavailable when you are in trouble, apart from becoming jealous and possessive about your partner. You start seeing this as inferior to be directed and shaped. For instance, you tend not to respect your partner as an equal. Because of this, you tend to act out on difficult psychological issues in acts of revenge, binges, and rages.

You need to resist the need for invalidating or dismissing the perspectives and experiences of others. Different people see life from different points of view. You need to learn to understand this fact. Additionally, you need to be aware that while you are being 'direct' at a person, you can also come across as intimidating unintentionally.

You need to make a habit of expressing your appreciation often and loud. On top of that, always avoid driving others to work as hard as you; not everyone is made in the same way. You need to remember that while sparring might be your style, it is not for others. And, you need to learn to negotiate and talk things out.

As type eight, you have a difficult time in lowering down your defensive stance in intimate relationships. One of your deepest fears is feeling emotionally vulnerable, which happens because of intimacy. Any sort of betrayal will prove to be intolerable and you will feel provoked to provide a powerful response. Intimate relationships are one of those areas where your control issues play out obviously and lack of trust becomes a pivotal thing.

You also have a sentimental side that you do not wish to show off to your partners, like fear of vulnerability. While trust does not come to you easily, when you indeed take someone into your sanctum, they become a stalwart friend and a steadfast ally. Your most powerful protective instincts are engaged when it comes to the defence of your family members and friends. You are also frequently generous to a fault committed by people whom you care about.

Anger

For most type eights, the anger stems from the feeling that you and your close ones have been treated unjustly. No matter how powerful you might seem, it is important for you to see how people would make you feel. You do not like being taken advantage of, betrayed, or bullied. When situations and plans do not go according to you, you tend to feel out of control, helpless, betrayed, and stuck.

Because type eights are in the Body Centre, you feel concerned with having and maintaining power and control. While other people might reflect on a particular situation because of shame or fear, you normally feel angry. In short, your type is an extension of God's power and protection, so it makes you very angry when you or others are treated unjustly or unfairly.

You mentally assess people as either strong or weak and treat them accordingly. This may lead to an all-or-nothing assessment of people and how you pay attention to them. All you want to know is the truth and you wholeheartedly dislike ambiguity.

You get comfortable if the truth is revealed, especially in the middle of a conflict scenario. The more information you have about the subject that is being discussed, the better you see the bigger picture. You do not like the idea of getting involved with a small piece of information about the picture.

Your mind will move towards your own needs and goals rather than those around you. You do not respond well to being charmed, bribed, or forced into doing something that you do not wish to do or that seems unimportant and dull.

The way you express your anger is similar to type sevens – you do it quickly. At times, your anger might manifest itself as detachment, verbally lashing out, and irritability. In most cases, it is quite intense. You mostly have a gym membership and know all the ways to vent out your need to spouting off physically. While you can keep this up, you also need to remember to introduce self-awareness to your practices. All your anger masks a bleeding and tender heart that you need to work on from time to time.

While it might go against the grain, you need to act with self-restraint. You tend to show your true self when you forbear your assertive nature to will with others. You can uplift and inspire people. You are on your best when you take charge of the situation and help everyone to come out of a crisis. While some might take advantage of you when you are caring for others, you will do more to secure the devotion and loyalty by showing the greatness that has been building in your heart, rather than displaying raw power.

While it might be difficult for you, you need to learn to yield to others at times. Mostly, the things that are on the stake are insignificant and you can allow others to run the show in their own way without having to sacrifice your real needs or power. If you have the desire to dominate everyone in the field, it is a sign that your ego has started to bloat – a signal of warning that you would likely face more serious conflicts with others in the future.

Most type eights tends to overvalue power. Having a position, wealth, or simply brute force makes them think they can get whatever they want and get away with it. You feel that it is important for you to be obeyed or feared. However, people who are with you for your power have no love for you nor respect you. While it typically might be looking bright on your side, you will ultimately have to pay the price of the power that you have collected, both emotionally and physically.

Mostly, type eights get angry at people and stay enraged from a previous action or comment. You need to express the reason then and there itself so that you do not burst out later. You might also get some consulting from your therapist or discuss the issue with a supportive friend.

Self-nurturing

You would normally express to stress by taking the challenges and problems head-on. You are assertive and bold in pushing for accomplishing your vision. However, this approach often leaves you overwhelmed and beleaguered. When the stress becomes too much to handle, you might suddenly switch your tactics and go into periods of isolation and retreat from the fight.

You are likely to pull back from the front lines to analyse the situation and place your strategies accordingly to regain control. You suddenly become isolated, secretive and quiet while trying to come up with ways to deal with the hassle. If the stress lasts for a long time, you will likely develop a cynical and cold attitude about yourself, other people, and life in a manner that is not considered healthy.

You need to learn to have fun with other people who will accept your non-conforming and outrageous behaviour. While at work, always ensure that you surround yourself with people who respect your honesty and direct approach.

You are prone to anger. When you are provoked severely or your personality gets unbalanced, your bouts of anger turns into a rage. You are naturally aggressive and can resort to violence when pushed to the limit. You can also be seen enjoying intimacy with people that you conceive as well and feel a bit compunction about stepping over anyone who is an obstacle on your way. You can be dangerous, brutal, and crude at your worst.

You need to be forthright and direct, yet open and flexible. You should learn how to stay present and steady whenever you are faced with conflict and confrontation; simply hold your ground. You need to call others on their threatening or destructive behaviour while being empathetic on what they might be facing. You should be able to express your own feelings as well, especially the tender and softer ones.

You need to remember that the world is not against you. Most people in your life care and look up to you. However, when you are in your fixation, it becomes difficult for them to understand you. You need to learn to let the affection in. Doing this will not portray you as weak, but will showcase the support and strength that you have in your life. You also need to remember that by believing in others and reacting to their words, you will mostly isolate them and realize your own fears. Learn to let people that are truly on your side know how important they are to you.

While type eights depend on no one and are quite self-reliant, the irony is that they depend on a lot of people. For instance, you might think that you are not dependable on your employees since you are the employer and they depend on you for the job. You think that you can dismiss them at any time and simply hire new people. You tend to think that everyone is expendable except for you. However, you are dependent that your employees do their jobs too, especially if your business can be managed on your own. But, if you alienate them, you will be forced to hire people who are untrustworthy and obsequious. Once you do, you will question their loyalty and fear of losing your position. Whether this situation takes place in your normal life or a professional one, your self-sufficiency is practically an illusion.

If you are a substance abuser, you can join a recovery program. Avoid having unrealistic expectations of yourself and find time to engage in activities that are physical and enjoyable.

Most type eights craves respect as opposed to being liked by a group or to have a status. You are always set to make an impact and never back down from anything. You pride in the truth, honesty, and strength. You also possess a strong inner sense of justice. You also value loyalty and remain devoted to the people who have proven themselves from time to time and will stand right by them until the very end. When you see that your loved ones are facing problems, you have the urge to stretch your wings to offer protection.

Things the type eights would never do

You would never eliminate all off-colour expressions from your vocabulary. You would also not state your opinions when you strongly disagree with the facts that are being stated. You would also never think of stepping down as the president simply because someone else can do a particular job better than you.

Even when you are playing a game of tennis, you would definitely try your hardest to win the game. However, you would also allow the opponent to let them do certain things in their own ways.

Positive things to tell yourself if you are a type 8

- I will try to show my loving and soft side to people whom I trust.
- All my great relationships will be worth all the small compromises and sacrifices.

***

Type Nine- the peacemaker

Overview

This personality type is usually modest, trusting and can be lazy. They are natural diplomats that seek peace at all costs and try to avoid conflict preferring stability. They concentrate on the environment around them and often forget their needs, for this reason they are quite easy to get along with. Moreover eights try to understand other people's perspectives, leaving them feeling supported. A nine's credibility is often questioned, as they try to put across the view point of various people, meaning that some could interpret this as being incoherent.

The negative side to trying to accommodate others can be when they begin to feel unsatisfied with it and start feeling overlooked or ignored. Eights hate being ignored, but at the same time will try to fit in and accept the situation, suppressing their anger. When it finally all becomes too much, they are prone to having passive aggressive behaviour patterns, which they are usually unaware of.

Basic fear for a nine type- Type nines act in this way because their basic common fear is of loss and separation from others.

Self-improvement for a type nine

People with type nine personalities, often referred to as the peace keepers, are highly adaptive, optimistic, and peaceful by nature. At the same time, they are afraid of separation and loss and their peaceful nature can sometimes be a defence mechanism for the conflict that they do not want to address. Let us take a look at some strategies that can allow you as a six to combat the negatives so the better angels of your nature can truly shine.

In Relationships

As a nine, your tendency to allow things to take their course can be truly detrimental to your relationships with people. Instead of hoping that things will change on their own, you need to take initiative and guide the relationship according to the needs of the people involved, yourself included. For this, you need to lay out your desires and needs and talk with your partner about the best way to go about addressing the needs of everyone.

Do not short-change yourself in this process or you will feel out of place and breed resentment and hatred towards them subconsciously.

When you're with a friend or a partner, don't just always go along with what the other wants to do. From time to time, ask them to join you in your interests. Not only will this allow your partner to know you better, but it will also make you appreciate them that much more. Make it a point to alternate between the activities that you want to do and what your partner wants to do. A healthy relationship is a partnership, and even though you may want to avoid any ensuing conflict, peace cannot exist on its own.

You are a good listener but you hardly ever get vocal about your own problems. This dynamic is too one-sided which may become pathological if left unchecked. Make it a point to ask your partner to hear you out, rather, ask them to remind you to talk to them about anything that you want to say. This will take care of any internal conflict that you might be going through.

There are certain phrases that you need to stop using. Some of these may include "I don't know" or "Whatever you want".

Instead, substitute them with more affirmative ones such as “I’ll let you know once I decide” or a simple yes or no. You have to be decisive in your speech for your actions to follow suit.

From time to time, whenever you feel like spending time alone let other people know. People cannot read your mind and need to be told what you want. Express your opinions and feelings freely. You don’t have to be the rock upon which anyone can lean. Instead, learn to speak the truth about your inner world and harmonize it with events in your environment.

Dealing with Anger

Type nines are known to be the most “out of touch” with their anger. As a nine, if you haven’t done much self development work, you may even deny that you experience anger. But the fact is, everyone feels it. Nines simply do not acknowledge it. This leads to the build-up of negative emotions to the point that a volcanic eruption of emotions can devastate anyone and everyone who happen to be in their vicinity.

In order to become aware of any signs of anger brewing within you, keep a journal and jot down all events and instances that lead you to feel bitterly towards someone or something. Soon, you will learn to find a pattern to your inner feelings and become more in tune with your feelings. This will allow you to become aware of your anger before you broadcast it to others in a thunderous manner.

There will be times in your dealings with people when you feel judgmental towards people. Pay close attention to what causes such feelings since more often than not it is either a cover up for anger or will lead to it. When things do seem out of line, don't revert back to your natural disposition of avoiding conflict. Voice out your concerns as best you can and do not act as if everything is merry when it is not.

Take up a meditation class and try to become mindful of the feelings that well up within you. At such classes, you will learn to feel the build-up of anger and stress in your body. This will allow you to deal with any negative emotion that may be brewing inside you as soon as possible and keep it from getting out of control.

Maintaining a work-life balance

As a nine, you generally have good work ethics. But sometimes, you can be plagued by procrastination which can make getting things done seem like an unnecessary chore. In order to combat this issue, try to have a daily routine and be consistent in your work. Instead of formulating huge goals, make a list of small chunks of things that you have to do in the day that will lead you towards that goal.

Give yourself realistic deadlines to meet and stick to them in earnest. Once you have completed the day's tasks, set the goals for tomorrow's task. This way you will ensure that you are being consistent in your work and aren't wasting time. Also, do not jump tasks. Finish tasks that are the hardest first when you are just starting out when your energy is high. Leave the easier tasks for later but don't forego them.

It is important that you ward off procrastination by doing things. Do not think too much about the ramifications or consequences at this stage. Simply do things to ensure that you get away from your laziness. Once you are in the flow, you can think about the ambivalences of your work and see how it fits into the big picture of your self-assigned goal.

To ensure that you are well motivated, reward yourself once your work is complete. This way the brain will associate work with said reward and you will feel good about your work. Additionally, ensure that the work you are doing is suited to your personality, your likes and dislikes. A work environment that allows you to contribute on a large scale and utilize your mediation skill is ideal for you.

Furthermore, learn time management skills to help you stay focused and on track. You can even lure yourself to do menial tasks that require your attention by listening to your favorite music while you work. This will ensure that you aren't bored out of your skull and still get the job done in a timely manner.

Getting over Procrastination and Making Decisions

As a nine, you tend to take longer than others to make up your mind, and thus procrastinate in the process. However, once things are sorted you can be quite like an unstoppable elephant. Therefore, it is essential that you clarify your goals before you start working. At the same time, go with your gut instinct. If, for the moment, contemplating on things seems like the right thing to do, then go ahead and give yourself room to think things through.

Once your goal is set, minimize distractions and let go of things that come in the way. Such elimination will further provide mental clarity that can be of great assistance when working towards your goals. As a naturally peaceful person, this will encourage you to relax while working since the conflicts, internal and external, will be taken care of.

To assist decisiveness, try making decisions on small things at first. Start out by making decisions based on what feels right to your senses – Do you like the color? How does it feel to the touch? Is it pleasing to behold? Your instincts will let you know if you are headed in the right direction. As you progress and making decisions start to become easier, work your way up to making big decisions.

As a nine, when matters become muddy or dense, you may become crippled to continue. But your greatest spiritual growth will come through reclaiming the sense of right action. Know that true growth is constituted in action and not in passive acceptance. Listen to your intuition and work your way through the mire of complexity and conflict since this is the only way to clear the path and fly towards your goal.

Developing Self-Esteem

For nines, one of the biggest challenges can be to overcome self-forgetting and improving self-esteem. As a peacemaker, you are a skilled mediator and work hard behind the scenes to ensure group harmony and flow. This, however, shouldn't come at the expense of your own needs and wants. Remember that the world won't be able to enjoy your gifts and talents unless you feel comfortable in your own skin.

Engage in a physical exercise program such as Tai Chi or other martial arts. Not only will it enable you to dispel your negative energy, the discipline and focus required to excel at such physical programs will trickle down to your inner life. The more equipped you are to handle the contingencies of life, the better prepared you will be when conflict arises. This will provide you with the wherewithal to handle disagreements and build self-esteem in the process.

Try not to distract yourself from your problems by indulging in excessive eating, binge watching TV, oversleeping, drugs, or other such negative outlets. These things will deal a heavy blow to your sense of self-worth once you are done with them. Instead, look at your problems from the perspective of an outsider and be objective about them. This will allow you to step back from the dissonance that you might be feeling and provide you with mental clarity to deal with the issues.

When a long-standing relationship ends, you may feel distraught and forsaken. As a nine, the fear that you might feel of being abandoned may be quelled by jumping into a new relationship. Refrain from doing so until you have had time to assess and understand what went wrong. Until unless you do so, you may simply end up repeating history and have another blasted relationship in your life. Take time out and contemplate for as long as it takes for you to understand and learn from your mistakes and weaknesses.

Even when you are in a relationship, take the time out to develop friendships and connections in addition to your partner. Not only will this broaden your emotional safety net, but it will also allow you to develop independence. Your self-esteem is directly tied to how independent you feel you are and this step is essential to consider since it will make you less dependent on your partner for all your emotional needs.

Sometimes it can be of utmost importance to simply ask your friend to lend you his/her ear and let things out. Ask your friend to not give any advice on the possible resolution of the problem. Simply talk and formulate your thoughts as best you can. This will clear away the mental fog that you might be experiencing due to the accumulation of negative emotions.

Doing things you normally wouldn't do

Though nines can readily forget their self to cater to other people to maintain a peaceful environment, it can bear heavily on them, especially when they aren't looking after themselves. Such habitual selflessness can lead to buildup of many negative behavioral patterns that need to be broken or bent from time to time. That is why, it is important to mix things up in the right amount to ensure that you, as a nine, feel safe and secure and up to any task that comes your way.

For instance, when your partner asks you to do something unpleasant in an authoritative manner, instead of contemplating or procrastinating, do what they ask straight away and without any forethought. This will ward off laziness and force you to do the task even when there is no clear indication of how to do it best.

Additionally, instead of staying peacefully oblivious of your surroundings, create uproar at your family gathering and be the source of conflict between people. This will let other people know that you are not as transparent as others assumed you were and will see you in a different light.

Tell your partner in advance that you are going someplace that your partner won't approve of so they can have the time to think about it and fight with you about it. Another thing that you could do to outrage people is to tell them that you are mad at them and tell them exactly what they did to make you angry.

When people want to talk to you, tell them that you don't want to listen or talk to them just to stir things up a bit. You could even confront people in a critical manner and rile things up instead of letting things slide.

Positive things to say to yourself

When things appear bleakest, remind yourself of positive things that you are capable of doing. Such words of self-encouragement can see you through the darkest times and get you back on track in no time.

- I won't shy away from asking for what I want and deserve.
- I will recognize my source of resentment and use it to energize myself.
- I will allow myself to feel guilty rather than rushing up to fulfil someone else's expectation of me.
- I have what it takes to do anything I want.

CHAPTER 4

PERSONALITY TYPES FOR THE HEAD CENTRE

Type five- the observer

Overview

This personality type is curious, inventive and logical. People of this type often become detached and isolate themselves from others, they typically enjoy their own company. However, they don't mind intervening in a conversation when the topic of discussion is in their line of work, as they like others to think that they know what they are talking about. They prefer not to participate in a conversation if they are not sure about what they are saying, in this case they would prefer to say nothing at all. People of this type are great observers and like asking questions in order to gain information on what they are trying to learn. They love to learn and are eager to possess knowledge, but this extreme passion for knowledge, hides an insecure personality worrying that they are not successful enough for this world. Fives are very objective and find it difficult at times to distinguish between thoughts and feelings. Their biggest fear is that of being incapable, they thrive for knowledge

Fives hate to depend on others and value independence, this aspect together with that of automatically detaching themselves causes them to lack in close friendships/relationships. Moreover, the relationships they do have are usually a struggle.

Basic fear for a five type- Type fives act in this way because their basic common fear is being helpless or inadequate.

Self-improvement for a type five

People with type five personality, also known as The Investigators, are high in perception, are secretive, innovative, insightful and curious. Though these strengths certainly do allow them to understand their environment and come up with ground-breaking insights, they can also have trouble with eccentricity, nihilism and isolation. Fortunately, there are various strategies that they can implement in order to counteract the negative side of their personality to allow the better angles of their nature to take over.

Doing things

If you are a type five personality, thinking about things comes easily to you. But sometimes, you can get so engrossed in it that you forego doing things. Instead of simply reverting back into your mind where you feel more capable, try to actively engage in doing things. This can be easier said than done but you can leverage your power of observation and learn from other people whom you admire and imitate them at first.

Through practice, you too can learn to put your foot down and do things without worrying about incompetence. Your thirst for knowledge is immense and you have varied interests. But in order to stop feeling overwhelmed or distracted by your innumerable interests, focus on one thing and develop yourself in it. This singular focus on one particular thing can keep you from getting lost in the complexities of your thoughts and lead you to certain mastery of the thing you are passionate about.

At the same time, your dedication and unwavering focus can lead you to become socially isolated. This has to be avoided if you want to get over your fear of the environment. Join a group where you are forced to collaborate with others, or pick up a team sport. This will dispel some of your negative energy and channel your thoughts productively. Although you may not realize this, but real understanding comes only through interaction and experience, not from knowledge alone.

As a type five personality, meditation can be worthwhile and particularly helpful in bringing your focus to the present moment instead of allowing it to flutter about. Psychotherapy or yoga can aid you as well, especially in communicating your feelings and nurturing your inner self. This can translate directly to becoming a better doer which can complement your intellectual prowess.

In Relationships

You may think that people are going to demand from you more than you can manage. But know that nobody is out to get you, and that you don't always have to hide behind what you know and can simply express your emotions as they come. Whenever you feel that you are withdrawing from your partner, tell yourself that you are going to stay connected for a "little longer". You have to allow people in your inner circle to know them completely and come to trust them.

Try not to overload your partner with your eclectic knowledge. Limit yourself to a few sentences if you are used to delivering long speeches on things that interest you. Determine if your partner is interested in talking about such things and, if they are, make it a conversation, not a monologue. Try to sense emotional cues that you get from your partner. Limit yourself to 3-5 sentences and check in with your partner.

If you feel the need to tell someone how much they matter to you, go ahead and say it. Similarly, if you want to do something for them, do it. Don't over think and start analysing the myriad of reasons why you should or shouldn't do so. People love it when someone says or does something for them, especially if that someone is a five.

At the same time, let people know of the boundaries that you have set and that they need to respect it. No one is going to read your mind; you have to say it to them in plain words. You could also become a member of a therapy group if you want to experience interacting with people. At such meetings, you don't necessarily have to speak, but when the need arises, the stage can be all yours.

Get out of your way and meet and talk to strangers on a daily basis. The more you do this the more you will be able to dispel your fear of people and become more comfortable in their presence. Know that other people too are opening up their selves to ridicule just as you are, and are equally vulnerable. Ensure that you engage with them in a respectful manner and are honest with them.

Find some common ground on which you can relate to other people, for this is the fastest way for you to feel comfortable around people. Don't just sit and talk; find a project to work on and get to know them through their work. You will be surprised at how easily you can make friends if you are not focusing on yourself or the strangers. The project has become a common goal, a catalyst that can allow you to make friends with ease and security.

At Work

You are highly inquisitive at work and a minimalist at heart which is reflected in how you maintain your surroundings. At the same time you prefer working alone and are fascinated by the unexplored realms of esoteric ideas. Therefore, it is extremely important for you to work in a field that allows you to explore and do your job without constant interruptions or emotional overloading by co-workers.

While at work, try not to hoard ideas and keep them from people. Your aversion to people may be well founded in your head, but you can help your peers understand and look at things from a different perspective. Ultimately, you have to get away from your myopic view and engage with people so as to address larger issues.

You should also get away from your perceived insecurities by asking others for help. Make it a point to ask for small help from people at your workplace daily. To your surprise, you will find that they do not pose a threat and are perfectly happy to help you. Also, try to find common ground with your peers and colleagues as this will allow you to work with them much more closely and make it easier for you to give them your time.

Your minimalist lifestyle is testament to the fact that you don't like to deal with the messiness of life. This can be a good thing as it makes you much more efficient at getting your work done than other people. At the same time, since synthesizing different things is your forte, you have to allow room for multiple things to come into your life, even if it is at once, to find the common thread that ties them together. Like a good puzzle, there are many patterns to be found here as well.

Your inexorable thirst for knowledge can actually be a cover for your deep-seated fear of being incompetent or a hack. In order to overcome this fear, engage with your craft on a level where you are open to criticism. Don't simply lock yourself away and do things on your own. Only when you open yourself up to analysis will you be able to know whether or not you need to work on yourself or you are just fine the way you are.

In trying to share information and expertise, you may come across as condescending to people. This can shut out people and make it difficult for them to understand you. To address this, make it a point to cut back on too much jargon and look into people's eyes when talking to them. This will allow you to understand the other person's perceived response and can dial back your haughtiness accordingly.

On Parenting

As a person with type five personality, the biggest challenge for you is to understand that the social needs of your child can differ greatly from your own. You also may prefer exploratory phase of childhood more than neediness of babyhood. But you have to understand that it is OK to not love the new-born phase and still be able to manage both your own and your baby's needs in a way that feels good.

To make the phase of babyhood easier, think of it like work – something you need to do as diligently as you can before you can retire to have some quiet adult time. The emotional aspect of being a father or a mother may be the most challenging. But being emotionally available for your baby can be the most uplifting thing that you can do for your own sensitive side. Unlike adults, babies are highly dependent on you and can be seen as anything but a threat.

When things do not go according to plan, ask your partner to help you with your child. However, do not become so isolated that you forego on your basic responsibility of being a parent. You have to assign pockets of time in which you have to serve your child's social needs, be it taking them to soccer practice, church, or some other social event.

There will be times when you will be able to impart your ideas and knowledge to your children. At such times, you have to resist the urge of dumping information on them. Instead, try to make it a game or a story wherein you give them pieces of the puzzle and allow their minds to roam freely and become curious about those things naturally.

You must work to ensure that you treat your children with kindness and learning that is appropriate to their age. As your child grows, you will find that you have much in common that you previously assumed. Parents with a type five personality tend to develop stronger bonds with their children once they grow up and develop their own interests.

Since type fives are more cerebral, the children may experience them sometimes as being emotionally withdrawn. Use your intuition and treat your child with kindness. Do not be authoritarian. Additionally, admire your child's ability to retain information and intellectual understanding since this may mean more to them, especially if it's coming from a Five.

Doing things you normally wouldn't do

The best way to avoid the pitfalls of the five personality is to ensure that you mix things up every now and then. This will not only keep things on an even keel but also allow you to enjoy your day to day life that much more. Let us take a look at some of the things that you can do to keep the negatives of your personality from taking over.

For instance, the next time you have the chance of exhibiting your extensive knowledge on some obscure field, give it a pass. This is a temptation that you should learn to avoid every now and then to ensure that you don't become too cocksure or haughty of your own intellect.

To get yourself out of your comfort zone and socializing, throw a big party. To make matters even more interesting, be the master of ceremonies. This will ensure that you are kept on your toes regarding your social endeavours. Through such practice, you will slowly understand what it takes to become socially successful.

Read through the societal page of the newspaper instead of going through the editorial. Sometimes, knowing what's happening in the lives of people can get you better leverage in your social life than going through the content-heavy journalism of the editorial pages.

If you want to get even more adventurous, try becoming a salesman and go door-to-door trying to perfect your sales pitch. This will force you to think of ways to influence people on the spot instead of thinking through every word before speaking. Though you may also be using the scientific names of things to help you remember them, make it a point to use everyday names to refer to things to give your mind a rest.

Finally, travel out of town with a group where you won't get time to yourself. Though being constantly bombarded with social information can tire you out, having nowhere else to go to will force you to find common grounds with people sooner. This type of stress is good for growth in the social aspects of life and can teach you a lot about yourself.

Positive things to say to yourself

Whenever you find yourself in a rut or feel that the negativity of life is bearing down upon you, tell yourself some positive things and bring yourself out from the down and out. This can be an important piece of information to have, especially when one isn't able to bring oneself to act. These are as follows:

- I must interact with people every day.
- Having genuine experience is necessary for true understanding.
- I don't have to be the smartest person.
- The world isn't out there to get me.
- I have full confidence in my abilities.
- I will not try to sound condescending to people.
- I will do things rather than stay in my head.

***

Type Six- the sceptic

Overview

People of this type are loyal, affectionate and responsible, but also anxiuos and suspicious. They thrive to belong and value security, in fact they are extremely devoted and trustworthy in relationships and at the same time prone to suffer from self doubt. They can be good problem solvers, even though when solutions to a problem seem to be too simple, they can be very sceptic and negative, questioning the solutions or advice put to them, sometimes even causing conflict with authority figures. Sixes are also known as fighters, they will not only fight for their own beliefs, but also defend the beliefs of others in order to help friends or family or even a good cause.

Their biggest fear is not having guidence or support, it is for this reason that they are very loyal to those around them. Sixes usually rely on external sources to keep them stable and happy, if these external networks are missing, they will easily create them and nuture them.
Behaviour

Sixes have a tendancey to dwell on things and allow worry to take over them, even when they are happy and positive they tend to not linger there. At their worst they are self destructive and irrational when they feel unsafe. They can also become unpredictable due to their internal confusion and blame others for their problems when they feel insecure.

Basic fear for a six type- Type sixes act in this way because their basic common fear is of being left with no guidance or support.

Self -improvement for a type six

People with type six personalities, or loyalists as they are sometimes referred to, are hard working and reliable people. They can foresee problems and promote cooperation, but at the same time can be very defensive and highly anxious. In order to minimize the negative aspects of their personality, there are many strategies that can be utilized, while the strengths are cultivated at the same time.

Building Self-confidence

In order to feel secure and feel supported by others, it is best to be around people who are trustworthy, accepting, and encouraging. Try to wean away from relationships in which people put you down or downplay your achievements. Your friends are your strong suit and cultivate life-long friendships as these will reflect directly on your sense of self-esteem.

There will be a time when you suspect if what people are saying about you is actually true, especially when it is a positive remark that they are making about you. Though a modicum of cynicism can go a long way in sifting truth from flattery, try not to overdo it as you will be depriving yourself of much needed confidence in yourself. Everyone, particularly people with type six personalities, requires some positivity from time to time, especially from their friends. So when you do get such approval from your friends, take it as it is. Disregarding it can leave you hollow and without content for the self to grow with.

Have faith in yourself, and know that you can overcome your fears and change whenever the need arises. Fear is the obstacle that you have to go beyond in order to grow. There is no way around it. So, do not be afraid of fear itself. Instead, see it as the opportunity that it is. Do not let fear paralyze you. Ask yourself – "Why am I feeling afraid?" The answer, more often than not, will be found in taking affirmative action instead of the anxiety that the body naturally feels.

Give yourself some leeway in your life; do not over think as to the validity of the path that you are on. There is no "right" way to live life. Work towards things that you like to do and as long as you are content, know that you *are* on the right path. Also, rewards yourself from time to time. It doesn't have to be an extravagant getaway. Even a little pat on the back for the things that you've done and the problems that you've overcome can mean a lot. Do not wait for others to tell you that you've done well.

Have a journal and write down positive things about yourself. Additionally, talk to yourself in a loving and nurturing way. You are your best friend and your worst enemy at the same time. So, make friends with yourself and do not allow yourself to get bogged down when things do not go according to plan. It is OK to make mistakes. Just because you couldn't keep your promise to a friend or a family member doesn't mean that it is the end of the world. No one is perfect, and you know it as well as anyone else.

On Relationships

Sixes habitually focus on potential hazards in a relationship. No wonder they are hyper vigilant and fear losing their support systems.

As a six, when you are stressed you may tend to overreact. Know that such reactions can have irreparable damage on your relationship. People will tend to assume the worst with such reactions.

Try to keep your emotions in check when you are stressed out the next time. Instead, be objective about the situation you are in. When in doubt, talk to your partner about it. If it is your partner that is causing you stress, try to look at yourself from a bird's eye view and choose a reaction that won't jeopardize your relationship for good.

Check your "projections" from time to time. It is possible that they are only partially true. There are many other areas that can provide you with important information to understand things from a holistic point of view.

You could even try to look at your own hyper vigilance in a humorous light. Every one of us has an aspect that, when seen objectively, is hilariously out of proportion. Your exaggerated watchfulness is a reaction to the fear of insecurity. Know that you have the choice to take things personally, or laugh things away. Whenever you feel such fear arising in regards to your relationship, simply observe it and calm yourself through deep breathing.

There are some other meditative techniques that you can employ in your life that can benefit your relationship with other people. Mindfulness is one such thing that you can do. When you walk outside, pause from time to time and notice the shapes, colours, sounds, and smells. Notice how it feels to take a pause and appreciate reality through your senses. Going into this appreciative mental state will allow you to feel thankful for the people you have in your life.

Maintain a work-life balance

Loyalists at work will generally stick around even after everyone has gone. Being disconnected from the group can bring up fears of your shortcomings. Such fluctuations in self-esteem can be curbed by nurturing one's inner self. Focus on your strengths to overcome fear. Acknowledge yourself for the hard worker that you are.

You have the tendency to become agitated and nervous when pressurized. Therefore, it is advisable to delegate tasks to other people when you are overwhelmed or overworked. The only thing you have to fear is fear itself.

That is why you may want to mentally prepare yourself for any contingency that may arise. Though this is prudent and wise, try not to think about the worst case scenarios and become overwhelmed to the point of paralysis. Instead, focus on the matters at hand, the ones that you are able to change, and let go of your worries.

At work, there may be times when you may become irritated by the slow pace of others. But refrain from over-reacting and becoming emotionally overstrained. Have patience with people. Know that they too are trying their best and have their own insecurities and flaws to overcome. Only by being cooperative and understanding will you be able to bring out the best in people and let go of your own worries.

At work, your loyalty can become a double-edged sword. Sixes tend to stand by a friend, a partner, or a job even though it may be time to move on. As a rule of thumb, you should analyse your position from multiple viewpoints, and take help from friends on the matter.

On Parenting

As a six, you have a natural tendency towards parenting. However, you can sometimes go overboard when it comes to protecting your children which can induce feelings of smothering in them. As soon as kids start to push and test your limits, take a step back and allow children to find their own space. Your personal fears regarding your children may be well founded, but children are naturally curious and should be given time to figure out their own sense of security.

As a parent you may want to ensure that your children's time is well utilized. But ensure that that doesn't substitute your child's need to develop his/her independence. Ultimately, you are working to make healthy individuals out of your children, and they need to make their own mistakes and learn from them. Whenever you feel anxious about letting your children make their own decisions, take a step back, breathe in deeply, and know that one day or another they will lead their own lives. All you can do is to equip them with the wherewithal to do so effectively and with honour.

Working through Anxiety and Fear

As a six, anxiety is your biggest nemesis. All your obstacles and defence mechanisms stem out of the need to get over your fears and anxieties. One way is to observe your fears without judging yourself too harshly for them. Try to be objective with your fears. Know the facts and analyse them from all angles whenever you feel you are overcome with anxiety.

There may be times when you may find it hard to make a decision and that can bear down upon you. At such moments, take a deep breath and tell yourself it is OK to stay in limbo, and to be indecisive. It is not going to be permanent and you will know what to do soon enough.

You can take up meditation or stress-reduction classes to work through your anxiety. These exercises and techniques can be of invaluable help to you in all areas of your life. For instance, when you are at home, visualize a peaceful scene that calms you down. Whenever you feel anxieties creeping back in your life, close your eyes, take deep breaths, and visualize that scene again. You will realize that soon your heart rate comes back to normal and you are feeling energized again.

Additionally, do not be a tyrant in your life. Cut yourself some slack when you've had a hard day's work and simply relax and do nothing. Do not call yourself lazy if you are not doing anything. Everyone deserves some time alone when they are not doing anything. You could even take up some physical exercise or a sport. Engaging in physical activities will allow your body to stay in prime condition and reduce the physical effects of anxiety.

Overcoming Pessimism

Due to your over anxious nature, you may find yourself rationalizing your fears through your pessimism. It is a common thought among sixes that needs to be addressed if they are to live their lives to their full potential. Though there is nothing wrong in the philosophy of pessimism itself, when combined with anxiety and fear, it can lead to disastrous results.

In order to combat pessimism, it is recommended that first of all, you get yourself into a daily routine. Your biological clock needs to sync with the daytime hours if you are to start cultivating positive thoughts. Additionally, you can read upon stoic philosophy of Marcus Aurelius or Seneca and follow in their footsteps to attain a firmer grasp of your actions, and become responsible for your actions

You may even take up a yoga course, meditate daily, or/and have regular sessions with a psychotherapist. These things in and of themselves can truly bring out the lighter side of your personality. Furthermore, in order to let go of fear, you need to focus on the tasks that you have at hand. This way, you can mitigate your overblown self-consciousness that is riddled with fear and anxiety, and allow yourself the room to focus on the things that you need to do. Slowly, as you come out of your shell, things will become much easier and you will grow by leaps and bounds.

Doing things you normally wouldn't do

It is good sometimes to mix things up and do things outside of your comfort zone. This will ensure that you don't get bogged down by the negative behavioural patterns of your personality. Let's check out some of them to see how you can benefit from them.

For instance, if you hear a familiar noise at night, simply go back to sleep knowing that it was the cat, and don't start worrying over whether or not it could be a burglar. Additionally, you don't have to feel bad if you end up being the cause of an accident. Forgive yourself for the things you did and move on.

If you are thinking of making a career change, try to forego taking advice from all your friends. Have faith in your own decision and go ahead with it. Similarly, when you go to a doctor, try not to ask for their credentials and have faith in them. If you have made a purchase, refrain from overly analysing whether or not it was worth it. Be confident in your choice and move forward.

As you get more and more confident within yourself, you wouldn't have to depend on other people to make your decisions for you. This will ensure that you remain safe and secure in your own decision making which will increase your self-worth.

Tell yourself Positive Things

When things seem bleak and life seems hard, remind yourself what you are worth and why you should keep moving forward. There is a light at the end of the tunnel, and the following statements should lead you towards it.

- I am strong and calm and have what it takes to get out of all troubles myself.
- I can do whatever I set my mind to.
- I am learning to trust my own decisions.
- It is OK to take risks and make mistakes.
- I will not overreact when I am stressed out.
- My reaction to things will be proportional to their intensity.
- The only thing I have to fear is fear itself.
- I am learning to let go of things that do not serve their purpose anymore.
- I will be patient with people and not get irritated by them.

***

Type Seven- the enthusiast

Overview

This personality type is fun, spontaneuos, optimistic and motivated. Sevens love socializing, meeting new people and experiencing new adventures. They value their freedom and enjoy exploring but do not like having limits .

Although they can be quick learner, they are very practical people and learn by practice not by theory or studying. This sometimes causes others to question their knowledge or credibility as they tend to learn quickly and overestimate to what extent they have absorbed the new information.

People of this type will often leave tasks that they are doing unfinished. They do this because their focus is on what they can experience or do next as they become bored quickly. This personality type likes to keep busy and are always active. Sometimes their listening skills are lacking because they tend to anticipate what people are going to say which means they lose concentration on what actually is being said.

Their biggest fear is of being in pain. The biggest negative aspect of this type is that they have lost touch with their internal guidance and compensate this by looking to the next adventure.

Basic fear for a seven type- Type sevens act in this way because their basic common fear is of deprivation and pain.

Self-improvement for a type seven

Also known as the adventurer, enneagram type seven are spontaneous, versatile, optimistic, and extroverted. They are also quite practical, high-spirited and playful. However, because these individuals are filled with talents, they tend to become undisciplined, scattered and over-extended. They are always on a lookout for exciting and new adventures but become exhausted and distracted while engaging in.

Mostly, sevens are seen with the problem of impulsiveness and impatience. When they are at their best, they will focus their goals on worthwhile objectives, thereby becoming satisfied, joyous, and appreciative. If you belong to type seven and facing the negative issues mentioned above, here are some tips that can help you improve your lifestyle and lead to development.

Health

Sevens are considered visionaries – they always manage to come up with new and adventurous ways to pleasure the world. What if you used this gift to visualize what you would look like if you are healthy? With the help of this vision, you will be able to work throughout your daily activities, for a long-term basis.

You also need to sit in stillness at times. Boredom get be a very frightful place for type sevens. You need to fight this boredom with food. Find the daily practices that will teach you about solitude like sitting alone with a journal, walking in nature, and working out. Your path to healing will start like this.

You also need to stop multitasking. Distraction is a big weakness for type sevens. While you might have a lot of great ideas, following up these ideas can prove to be a struggle for you. You need to look for a single way to reach your goals and develop a more disciplined lifestyle. You need to stop rushing into things and layer your thoughts. Take time to slow down and do just a single thing with your full and undivided attention. This way, your mind will remain strong and you will also get things done on time.

You will need to cultivate some healthy habits of exercise, sleeping, and eating. This needed to be said because some sevens tend to go to the extreme and neglect their health. You can opt for an exercise program like tai chi or swimming for instance.

You need to be careful so that you do not spend, drink, or eat excessively, especially when you are stressed. These will amplify your stress and fear.

Stress

You need to recognize your impulsiveness. By observing your impulses rather than giving in to them, you will become a better judge of what things are worth acting on. The more you can resist your impulses, you will be able to focus your attention and mind into things that will do you good. In return, your mind will be clear and focused, thereby minimizing stress, tensions, and worries.

You can consult your friend that will bring you back to the realities of life and offer practical solutions. This is why you need to keep your friends close to you. You should also strive to break down your important goals into smaller steps that they can be tackled a lot faster. You need to list down areas that you can improve upon and share it with a friend that can help you go through these points with conviction. You can also dive deep into subjects like religion, ethics, anthropology, sociology, psychology, and philosophy to learn about your values and inspire others to find their happiness with the help of endless positive energy.

At times, the world might not offer you what you would have wanted. In this case, it is always better that you be grateful for what you get and simply focus on the coming future. You need to be appreciative of your ideas and stories and avoid being too negative or adamant of doing something in your way only. Take responsibility for your actions, while staying positive at the same time. You need to stop talking and start listening. Let people know that you are there for them.

You need to encourage your sobriety by staying grounded, balance out good ideas, and build common sense. Stress on feedback and be present for your friends when they need you.

You should never tune out your problem in the hopes that they will simply vanish. In these cases, you need to look for a friend or a professional counsellor so that you can talk and not allow the stress to build in your head. You just need to accept your feelings, no matter what they are. Once you do, the stress will pass and not bother you again.

As a type seven, you are quick-minded, flexible, spontaneous, playful, and curious. You tend to see the bigger picture. You are expected to entertain your friends because you are fun; however, it might feel like a burden at times. If you are stressing out, you need to stay busy and start having some fun. The seven inside of you will always try to escape from the painful and unpleasant things in life. You might also get bored easily, so you will often face trouble completing your tasks.

Relationships

If you happen to be in an intimate relationship, how much time you and your partner spend together and away might possibly present a problem. In this case, you need to reach to a conclusion as soon as possible and the amount of space and time together that you both need from each other.

As mentioned in the above sections, type sevens are high-spirited, playful, extraverted and busy. However, their basic fear is feeling too much emotional pain and not having any sort of fun. You wish to be happy and believe that life is all about fun and having a good time. Experiencing something like difficult emotions is something that you would avoid actively.

According to several studies, type sevens feel minimum compassionate care and a lot of perspective-taking. This fact is quite contradictory, especially judging by their understanding of the situations and feelings of others. Their tendency to think instead of feeling is the reason why they are considered so low on empathy.

If you are a type seven, you might turn out to be insensitive towards the needs of others while they are trying to have fun and avoid difficult anxieties. You are more focused on making plans for the future and your desire to achieve your targets prevents you from expressing or feeling any sort of empathy towards other people.

As said above, you avoid feeling empathy because you think that it interferes with your goals. However, if you manage to develop awareness for your behavior and type, you will be able to see the gaps in your emotional and social development and find different ways to connect the bridges. Once you can do that, you will see that it has become easier for you to forge stronger connections along with people who are important to you and play an important role in your happiness and survival.

You need to stop avoiding your emotions, slow down, and take your time to reflect on your feelings and experiences. Being constantly excited and on-the-go can increase your anxiety and fear that you might have to face something ugly and unpleasant. However, if you manage to face your fears, you will gain some security.

You need to recognize your spontaneity and not opt for hasty action. The next time you feel the urge to charge head-on to something, you need to slow down and take a moment to think whether the action you are about to take is worth it. You need to learn to resist your impulsive nature and feel your values to see what is really important here.

You also need to learn how to listen to people. If you are quiet, you will be able to understand your loved ones and learn about them, including their needs and their feelings. Stop rushing around and appreciate the world in solitude and silence; this will help you reconnect yourself with your feelings and give you a deeper understanding of the person that you really are. Once you do this, you will be able to connect with your loved ones more authentically.

You also need to learn to slow down. While experiencing new things in life can be quite fun, you also need to devote your attention to appreciating the moment. If you focus on things that are coming in the future, you will miss out on the present and will be trapped in the cycle of wanting more. Never fail to be satisfied with what you have right now.

Valuing quality over quantity is something that you need to engage in. While it might sound like a cliché, spending a good amount of time with the people that matter will help you develop love and empathy towards them. If your relationships are based on quantity, you will have a lot of people around you. However, none of them will be able to feel closely attached to you and you will have difficulty feeling empathy for them.

When it comes to intimacy, you need to set a designated time aside for that. You need to be sensitive and tactful. This way, you will be able to see things from the perspective of your partner and understand what they are trying to convey. If you are open to hearing about feedback about your traits, you will receive a lot of scope for improvement. For instance, if someone says 'you did not ask about my day', it would mean that the person wishes to talk about how the day went by for him/her.

You should strive to answer to the point. No one likes a self-centred person. Answer to only what they want to hear, nothing more and nothing less.

Work

When it comes to work, type sevens are hare-brained, impulsive, self-centred, and uncommitted at their worst. You tend to jump at any random opportunity and struggle evidently with the follow-through. You are also known to overload your plates and get sloppy when things around you start getting fast-paced. While you can come across as arrogant, you are, in reality, fragile and will do almost anything to dodge criticism.

You need to improve your listening skills. There are people who are always trying to teach you something that you might not know. However, you think that you are smart. If you wish to improve your listening skills, you will be able to provide space for what other people have to say. You should listen to them earnestly and genuinely. You should never rush to provide a back answer or break the conversation to share some personal anecdotes. You will be amazed to see how much you are loving the conversation when you are not anticipating any discussion from your side.

You also need to realize that you will not have everything available at the moment right now. You would want things would be better if they happened now. But, as the moral goes, the best things in life will take its own merry time. Most great opportunities will come back later at an even better time. However, you need to dive deep down and look into what is best for you currently. Mostly, your life is already doing great. You need to live it rather than planning it out. Let things run its natural course.

When the option arises, you should always choose to work for yourself. Additionally, you should be considerate to others and not expect to work with the same levels of proficiency and effectiveness as you. If they cannot keep up with your fast pace, you should consider slowing down. Slowing down will help you look at things from a different point of view; you will be able to understand things from a unique perspective.

Whatever work you are doing, make sure that you are doing it with your full concentration. Let nothing else distract you from the task at hand. You should remember whatever effort you put in your work will surely reward you and result in great satisfaction. Most sevens tend to think that if the work does not get the pleasure, it is not worth the time and energy. Always look out for a career where you can put your ideas into action.

Things sevens would never think of doing

Complete reading all books before grabbing hold of another one. Also, they would never jot down a to-do list for the entire week and following them as planned.

Telling stories and incidents without exaggeration and spend half a year at a Zen center in a completely silent meditation mode. They would spend an entire day with a friend and not suggesting a single short-cut to help get the job done faster.

They would also never volunteer to work on a hotline for grief or spend almost the entire day lending an ear to a friend so that he/she can upload their problems and talk for reducing the load.

Positive things that you can tell yourself

- Clarity and truth are in sight when the dark is balanced with the light.
- There is no need for anything more. I have enough and there is enough.

CHAPTER 5

Personality Types For The Heart Centre

Type two- the giver

Overview

People of this enneagram type are caring, kind but also possessive. They feel the need to be appreciated by the people around them, so their principle focus is that of building relationships and doing their absolute best to help and support their friends and family. In fact, people find it easy to be around this type as they come across as warm and generous and have a tendency to sacrifice their own needs and time to be present for others who need it.

The biggest fear for a two type is to be unwanted, they have the need to feel loved and appreciated. This constant will to be accepted by others is why they are so concerned in pleasing people; this could be seen as manipulating people into making them show their appreciation. However, twos do not realise this normally, they only tend to see the positive side to this personality trait.

Being so generous towards others does make twos an easy target for people to take advantage of them, and when this happens, it triggers the emotion of anger and resentment in the people of this type, leaving the person at the receiving end of the anger taken aback as they do not expect it from this personality type. They also prone to not being connected to their own needs because they put all of their focus on other people's problems.

Basic fear for a two type- Type twos act in this way because their basic common fear is being unloved or unwanted.

Self-improvement for Type Two

People with type two personality are known to be highly altruistic and empathetic. Although they have their strengths, their weaknesses can take over their lives easily too. Fortunately, there are many strategies and behaviours that they can employ to ensure that they do not get sucked into the negative patterns of their personality.

Improving Self-Esteem

Helpers, as people with type two personality are commonly known, can have problems asserting themselves or feeling sufficient within their own skin. In order to combat this, the best thing that you can do is to do things just for yourself. When you don't involve other people and do not seek their approval, you learn to nurture yourself and become satisfied with who you are.

Learning to meditate is another easy way by which you can stop worrying about all the ways in which you feel insufficient and bring your focus upon yourself. Through this, you will get in touch with your inner child and alleviate some of the hurt that s/he may have felt because of neglect. Talk to yourself lovingly and assure yourself that you are whole and complete just the way you are.

Another way in which you can boost your self-esteem is by asking others for help. Helpers want to help others but rarely ask for help. They feel as if they are not worthy enough to ask others to assist them and are thus afraid to even approach other people. This has severe consequences on their sense of self which can only be repaired through asking for help. Be bold, but not overbearing when doing so and you will find that you are just as deserving of help as anybody else.

Additionally, you can seek help from a counselor and learn how to discuss your own problems. A professional can really bring you to open up about yourself to other people which can be carried forward in your daily interaction with people in your life. As mentioned earlier, you don't necessarily have to be the healer of the group in order to feel satisfied. Your fun loving nature is good enough for people to like your company.

Develop Assertiveness

Though it can be hard to say no to people, make it a point to do so especially when your own needs aren't being met. This becomes even more important when you realize others' problems are beyond your ability to resolve and may end up stressing you out.

Have personal boundaries in your relationship with people. If you allow people to encroach upon your personal space, you may end up resenting them and feel that they are taking advantage of you. Such feelings need to be nipped in the bud because if you allow them to fester and grow, they can really make you hate people in your life and become a pathway to self destruction.

As a type two personality, repressing anger and negative emotions as a defence mechanism comes naturally. But such ways of dealing with inadequacies and perceived hurt can manifest itself in a myriad of ways. The best way to understand what internal obstacles are depriving you from mental clarity is to write things down. Start a journal and write down in the most honest way possible all your resentments and bitterness. This way you will develop a relationship with your shadow-side and understand the things in you that lead to become emotionally distressed.

Also, learn to speak up when you feel you are being treated unfairly. People may have gotten so used to seeing you as the quiet helper who can take on all gibes and misgivings that the only way to ensure that they don't take further advantage of you is by voicing out the truth. Do it as soon as you feel it and in the most reasonable way possible. Not only will you gain respect in other people's eyes, but you will also develop assertiveness that will quash all feelings of resentment and anger.

Relationship Goals

People with type two personality excel at empathizing with people and forming life-long connections. But this sometimes comes at the cost of one's own inner life. Before you attend to someone else's emotions, try to understand what your own needs are. Though gaining the approval of your partner can have its own advantages, it isn't a good substitute for being loved for what one is like.

Being an emotional sponge, you have to be very careful what you absorb from other people. Having personal boundaries can mitigate this to some extent. It can be a good exercise to reflect on some of the ways in which avoiding your own needs leads to problems in your relationships. Now compare this with a time when you took good care of yourself and how the positive emotions within you had a positive impact on your relationship.

Choose relationships in which your own authentic self is appreciated, instead of being chalked up as a mindless meddler. Do not offer advice and help when it is not called for. Wait until you are asked for it and it will be celebrated all the more. Moreover, try not to over give as this be overwhelming for people.

On the other hand, accept help graciously when people give it to you.

Though you may not realize it, being a helper may also mean that you may be nurturing feelings of entitlement in your relationship. And why wouldn't you? Haven't you done enough for your partner? But it is not prudent to confuse entitlement with pride. The best way to address this issue is by having an honest conversation with your partner about your feelings and your need for appreciation. You will ultimately have to understand that other people, not even your partner, will appreciate you if you don't appreciate yourself. However, if you feel helping is an inextricable part of your nature then go ahead. But if you are expecting something in return, it isn't helping, it's business.

You can further nurture yourself by spending some time alone and developing your inner self through activities like meditation and yoga. Try to refrain from being on the computer or your phone and simply be with yourself. This will broaden your horizons and teach you to focus on the moment rather than on the negative aspects of your personality and your prior misgivings. The advantages that you get out of this will directly be reflected in your intimate life with your partner.

On Parenting

No parent wants their child to get hurt or be bogged down with the tragedies of life. But at the same time you ought not to shower hugs and treats to your children every time they are upset. Every parent cares for their children, but you need to maintain a fine balance between loving your children and allowing them to nurture their own independence.

It can be hard for parents, especially those with type two personality, to refrain from protecting their children and providing love and affection. However, this same love, particularly for certain type of kids, can feel like smothering. Children learn from disappointments and pain and you need to give them the space to understand and reflect on the things that went wrong.

In order to stop constantly worrying about the needs of your children, develop some interests of your own while they are away at school or at work so you wouldn't constantly bombard them with text messages and calls (especially when they are of age and are living their own lives).

Take up work at your local community center where help is urgently required and appreciated, or learn a skill that can allow you to help others in different ways. This is not to say that your children do not need to feel valued by their parents, but there is a time and place for such things and understanding when help is called for is crucial.

Furthermore, do not instil guilt in your children. Children, especially when they are below the age of 10, learn through osmosis and will take on your own patterns of behaviour and dealing with things. Though your helping nature can be of great value, ensure that you are teaching them to be dependent on the approval of others.

Similarly, do not over engage your children. They too need some time alone just as you do to nurture their inner selves. Being there for them every single time or giving into their every whim and desire can make them feel entitled and spoil them for good. Have an authoritative parenting instead of a complacent one. Set rules and boundaries to let them know what is expected of them and do not be permissive of everything in the name of parental love. In the long run your children will thank you for it.

Developing Healthy Co-dependency

In order to develop co-dependency, you need to ensure that you know the nature of the person you are getting into a relationship with. It can be easy for you to jump in and help the other person better themselves through your help. But refrain from doing so in the initial stages and go slow. Get all the information you can about the person through interacting with them and talking to people who know them. Through all this, be as objective as you can or you may end up being used for your virtues.

Accept only those people as friends or partners who are equal to you. Do not jump into a relationship with someone who is excessively needy or unavailable. Not only will they not appreciate your time and effort, but they will also bring your self-esteem down and hurt you in the process. It is not your burden to rescue people, especially those who don't want to be rescued. Everyone is responsible for their own individual behaviour.

If you have had a traumatic event in your life, such as a broken relationship, you may find it easy to find comfort in nostalgia and hold on to sentimental photographs, mementos, journal entries, etc. Such times can make you debate whether or not to hold on to the past, or to get rid of emotionally taxing feelings. The only way to heal as such is to accept what was and forgive.

At the end of a relationship, resist the urge to jump into another relationship, especially if you haven't understood what led to the dissolution of the previous one. Instead, spend some time by yourself and develop new interests, relax with friends, and get back in touch with your inner self. Be in the company of people who want the best for you and don't always expect things of you.

Maintaining a work-life balance

As a type two personality, your biggest strength at the workplace is having good relationships with your peers. This will allow you to thrive if you can manage certain obstacles at work. Though you genuinely like to help people achieve their best, be on the lookout for people who only want to get their work done through you. Therefore, it is essential that you maintain limits and not take on unnecessary work just to please other people.

It may not come naturally but you have to establish some assertiveness when dealing with people and being objective about your task. Put your foot down and say no when someone wants you to do their work for them. Not only will you be saving yourself hours of unappreciated work, but you will actually be allowing the other person to grow through their work. Ultimately, you will gain their respect and even their friendship over the long run.

Finally, ensure that your job is cut out for you and is suited to your personality and your interests. Ask yourself what you can bring to the table and learn the ropes as you go. It is best to be in a field where interaction with people and building strong, working relationships are appreciated and respected. These things come naturally to you, so take advantage of the fact and you will find your work fulfilling.

Doing things you normally wouldn't do

Sometimes, when you find yourself being bogged down or overcome with negative sentiments, it is best to shake things up and to do things that you normally wouldn't do. For instance, try to not greet or smile at people for a good long week, or even a month.

Additionally, when complimented, simply say "thank you" and move on. Don't brush it off, discount it, or brood over why you don't deserve it. Just let it be.

Don't make excuses when you aren't able or willing to do something for someone else. Furthermore, don't return a favour, or even when you do, don't do more than what is required. Ignore your co-worker's indifference to you and be cold to people who don't care for you. You can't please everyone all the time, so stop trying.

These things may appear harsh or totally outside of your nature. But every once in a while we have to assume a different role to break away from our pathological behavioural patterns, so that when we come back to ourselves we can respect ourselves that much better.

***

Type three- the achiever

Overview

This personality type is image conscious, energetic and ambitious; they invest their energy into becoming successful and usually succeed at it. People of this type love setting themselves goals in order to constantly improve on their skills and capacities, as a result people around them tend to look up to them or even use them as their role model. When threes are in a positive happy place, they are usually self-confident and

A threes' biggest fear is of being worthless, they thrive on the attention, which surrounds their success, and become afraid when they begin to lose this feeling of accomplishment. They sometimes tend to become so involved in being successful and give so much importance to their external image (which could be their family, their beautiful house and their big car) that they lose track of who they really are; of their true essence.

People of this type usually find it difficult to talk about negative or difficult situations, and when this type of topic pops up, they tend to avoid it or dismiss it and cut the conversation short.

Although people usually look up to threes, they can sometimes be seen as impatient and dismissive due to them being distracted and concentrating solely on their goals. A similar mechanism is also true for their feelings; they don't like their feelings to get in the way of their success, so they often put their feelings aside and ignore them.

Basic fear for a three type- Type threes act in this way because their basic common fear is being unaccomplished or unworthy.

Self-improvement type three

Type three people generally have the fear of being worthless and not making the most of their time. They are too status and image conscious, which is why they put on different facades in front of other people. Moreover, they need validation from others almost in everything. So, if you are a type three and relate to all these characteristics, you can be a better version of yourself by doing the following things in order to improve your personal and professional life.

Self-nurturing and relaxation

The hardest work is to go idle. - Jewish proverb

Slow down your pace. This way you will be able to see and observe the world around you as you slow down. You need to take out time for meditation, resting, and other stress-relieving techniques like saunas, steam baths, and massages. If not, the stress of overworking and achieving your goals will wear you down and lead to a lot of problems, including physical and emotional ones.

Most three's feel that they cannot meditate. Hence, if you plan on working something slow-paced, you need to be more mindful of your surroundings and simply start off slow. This way you will be able to see and appreciate things around you.

Even if certain feelings are not as shiny as you think, you need to welcome them normally. You need to make time for doing other activities that you value apart from your work. These activities will keep you energized and fresh.

Furthermore, you need to understand that failure is a part of the journey and not the end of the world. It is natural for you to place more attention on the 'winners'. However, you need to remind yourself that life is much more than just about medals and first prizes; there is more than what meets the eye.

You need to feel and experience love as a human being rather than all your accomplishments. You can reduce your stress levels by simply accepting and appreciating your current success levels.

You need to try and accept all your other preferences and desires, which can help you become more aware of the real world. And, you must learn to listen without having any other agenda. Be curious and know what the other person is trying to say. Also, be mindful of how you are finishing or interrupting their sentences.

Additionally, you should practice self-forgiveness and self-compassion, especially in situations where you feel like a failure. Also, make sure you pack your bags and leave for a long vacation at least once a year. And while you're at it, remember not to carry your work.

Work

You need to realize that there is more to you than just your achievements. Your success will make you worthy of admiration and respect only for a short time but connecting with the right people will validate you in the long run.

Your confidence will mostly be projected as certainty. While in most cases, certainty might seem like a positive thing, it can make you a lot dismissive to alternative perspectives, thereby keeping the views of other people out of the picture. This might overall decrease their openness to your goals, plans, and inputs. You must ensure that you always keep your ears open and listen carefully to others when they are talking about their perspectives.

You need to realize that people are not always as efficient as you are when it comes to professional work. Always remember that people might not have your ability to stay focused or emit the same levels of energy. While choosing jobs, you need to look for something that satisfies your inner desires. At times, type three's tend to choose careers that do not cater to these desires and needs.

You need to develop social awareness. Most type three's reach success, but at the cost of getting involved in projects that will have nothing to do with personal advancement. Working towards a goal where your personal interests lie is a great way to finding your true identity and value.

Always compete with yourself. Always strive to be better than you were yesterday. Do not compare yourself with other people as there will always be someone more successful, richer, and better than you. But, before you go ahead and forge your own path of success, take out a second to consider what other works will be negatively impacted. Get busy with other work while you are waiting for your chance. You need to be aware of the decisions you make in a hurry; you might potentially be ending the career of a colleague this way.

Develop ties and relationships with genuine people. You should remember that true happiness does not come from success or a perfect image. It only comes from people who truly accept the person that you are, with all your negative aspects as well. You need to be supportive and make efforts to appreciate the work and contributions done by others.

You should remember not to push yourself too hard, especially at work. Take your time out occasionally to unwind and enjoy the simple things in life. Self-development and ambition are great qualities, but driving yourself to exhaustion is not the way to do it. Take breaks and temper them with rest periods to reconnect with yourself spiritually. At times, simple breathing is enough to improve your outlook.

Relationships

You need to ask yourself about what and how you are feeling. You should allow real emotions to emerge rather than trap them inside, which can hamper your production.

Learn to slow down, moderate your pace and detach from goals and performances temporarily. You should always try to express your happiness and appreciation to your partner. At times, type three's tend to think of themselves as the superior one of the couple since they have accomplished a lot.

Whenever you are writing down your weekly schedule, you need to factor in the time that you will be spending with your family and friends. Set your personal boundaries and limits on work and make time for your loved ones. Whenever one of your friends or family members are facing any sort of trouble or problems, you need to listen to them. While doing this, ensure that you do not provide any sort of advice, unless they ask for it. Your friend/family member might simply be asking for a sympathetic ear.

Apart from setting up success and expectation standards for others, you need to look inward for and towards your own identity. Develop cooperation and charity in your relationships. This can be done by pausing on a busy day to connect with your loved ones. You do not need anything spectacular about this – just a few seconds of pure appreciation. In doing so, you become a more loving person and a faithful friend. You need to realize that love comes from being you, not having or doing.

In an attempt to be accepted by others, some type three's tend to lose touch with their real selves. You need to resist what you are doing and just accept the way that you are. Develop your own unique character by investing time in discovering your own core values. Opt for volunteer work whenever you can, where you give something to a needy person with a pure heart and always for the sake of giving away.

Ask your friends to nudge you whenever you are fudging the truth or being inauthentic to impress someone or to make yourself look good; ask yourself why are you being this way and are you really going with your image. Whenever you are bombarded with criticism, always look for grains of truth instead of arguing back. Learn to listen to people and be receptive to it.

Always project yourself as the person that you really are; do not change your personality just for winning over them. Make notes of your undesirable traits and work to improve them slowly.

You should learn to value connection and empathy highly. You have to be more patient when it comes to dealing with your own feelings and others'. Ask yourself what actually matters in your heart; ensure that you make time to pursue it.

For parents

As parents, you need to list down all the accomplishments of your children. Begin by how far you have come as individuals and as a family. Reflect your accomplishments to your children to give them a better perspective. Share your goals with your kids. Let them know what you want to chase down this year and how you plan on achieving them. Your children will realize that self-improvement is a long and ongoing process.

You need to be aware of what type of expectations you are setting up for your children. If the pressure seems too much, you need to avoid since it can create a lot of emotional problems for them in the later stages of life. Put them on a positive spin and keep the goal-setting sessions positive with the children. For example, instead of using terms like 'get better at' or 'lose weight', you need to say things like 'learn more about' and 'be more physically active'.

Keep the list short and sweet. Jotting down a lot of goals and fulfilling them is nearly impossible. There simply isn't enough time in a single day to do everything. Ask your children to aim for goals that can be tackled in a short time. Discuss and trim down ideas and aspirations. Always allow them to pick goals on their own. While they might ask you for guidance while coming up with goals, you need to ensure that the end results are all their ideas.

Set goals for your entire family. While individual goals are important, your family as a whole matter as well. Pledge to spend more time together and learn or achieve something new every year.

Feelings

Always speak the truth. Find honestly within yourself and let others know how you truly feel and the things that you need. There is no need to hide who you really are. Always find and stand up for your values. Most type three's tend to do what they 'should' do just to meet the expectations of others. They do not try to find what they value the most and will do just about anything to be accepted.

You need to become aware of the differences between your real feelings and the feelings that you tend to fake because they seem appropriate for that time-being. When anxiety starts to arise, you might tend to jump into doing some activity. In this case, you need to sit down and compose yourself. Take a walk and refresh your mind.

You need to calm down and stop becoming so self-deceptive. You need to believe in your public image or you will lose touch with who you really are and create a lot of confusion. People will start seeing you as uncaring, opportunistic, and insincere. While focusing on your goals is important, it is also equally crucial that you show your appreciation for your cared ones. Asking about their day and listening helps build bridges and keeps the empathy alive.

Look out for people with whom you feel compatible and loving. It is a relationship, not a competition. You do not need to impress anyone with success or status. Being authentic is much more impressive than bragging.

It is always difficult to discuss negative issues and one would mostly dismiss or rush such conversations. This case will seem particularly true if the criticism points towards the shortcomings and mistakes. Listen to them carefully and start your work towards improving them. It is always okay to express the vulnerable side of you – hurt or disappointed.

Always make sure that you reach out to others. Find ways to engage in people with work or interests that are outside your comfort zone. This way, you will build connections and learn the value of working together.

Things type three would never do

The type three's would never dare to make a list of goals to achieve for the coming next week. Instead, they will take things at a normal pace. They do not take charge of the meeting if it is being conducted ineffectively or inefficiently. They will immediately want to get it back in order. If they feel that a particular job needs some doing, they will do it immediately. If not, they will at least make a note of it. They would rather not sit at a silent meditation retreat for a week.

They will never brag about their achievements at any type of school or college reunion. They prefer to let their work speak for themselves, instead of having to blabber out everything they have achieved. Type three's will refrain from making others uncomfortable, especially on the ones that they are interested in.

Positive things to say to yourself

- I need to remember that God has made me in His image and wishes for me to keep continuing. He made me a kind and talented person.
- I will give my feelings the same level of priority and importance as my accomplishments.
- The best thing to do right now is to take some time off from work, relax, reflect on life, and grow as a human being.
- I must practice solitude and learn to value and perform for God. Being alone in His presence will help me become gracious.
- The measure of my worth will only be done by my standards, not anyone else's.
- I need to spend time and decide who I want to be. I need to stop chasing after opportunities and positions that are not meant for me. I will work with all my heart and bring enthusiasm to the lives of people who matter to me.

***

Type four - the individualist

Overview

This personality type is creative, reserved and expressive, they are also known as romantics. They are usually artistic and are attracted to beauty. People of this type are not afraid of being different, they actually feel the need to be seen as authentic. Fours are quite emotional and are aware of not only their feelings but also of the feelings of the people around them. Thus, they can be seen as being very sensitive.

Fours do not mind sharing personal experiences and open about their feelings. They don't want to be accepted by everyone, because they enjoy feeling like they are not like everybody else.

As fours often feel misunderstood, they look for a relationship with someone who they believe understands them and work at cultivating this relationship.

One of their biggest fears is of not being significant, which is why they spend their energy on searching to create a stable identity, as they are prone to having the feeling that something is missing. When they are not at their best, they could suffer from low self-esteem.

Basic fear for a four type- Type fours act in this way because their basic common fear is not having a unique identity.

Self-improvement for a type four

People under enneagram type four are reserved, sensitive, and self-aware. They are also personal, creative, and honest; the downside here is that they can be very self-conscious and moody. They will often withhold themselves from others due to the feeling of being defective and vulnerable. Type four also has a feeling of exempt and disdain from the normal ways of living.

While at their best, they can be highly creative, inspired and capable of transforming their experiences; but they can also typically showcase problems like self-pity, self-indulgence, and melancholy. If you are type four and facing these issues, here are some things that you can do that will help you develop your overall life.

Self – esteem

Do not pay a majority of your attention to your feelings; they are not exactly a true source of support for you. You should never forget that your feelings will tell you something about your own self at the current moment, not anything more than that. You should be proud of all that you accomplish with your special talents and gifts.

You should stay humble. We are all made in unique ways and are gifted with different talents for a reason. If you learn to be open-minded consciously and tolerant of others, you will appreciate the subtle differences between people that make us human. You can read books on sociology, anthropology, history, and philosophy that will help you open your eyes to new perspectives and broaden your scope.

You should always treat yourself with feelings and needs that you were not able to get in your childhood. It is okay to treat yourself to some compassion and love once in a while. The past is gone and the future remains to be seen. Hence, you need to think and value whatever you are doing right now.

When it comes to creativity, you need to commit yourself to recreational work that will bring out your best personality. You need to find ways and methods of doing everyday things in a more playful and creative manner. If it comes to it, try hard to make it your earning source as well. However, if you are unable to make a career out of something that you are passionate about, then you need to ensure that you get the time for practicing your art so that it can grow side-by-side.

You need to devote your heart to mastering self-discipline. Self-discipline comes in several different forms and has a strengthening and cumulative effect. On the other hand, fantasizing, sleeping, drugs, alcohol, or excessive sexual experiences will have a debilitating effect on your mind.

Notice the type of behavioural traits that you like and dislike in a person. Cultivate and work on the ones you like. You can also choose to serve as an inspiration to others.

Relationships

You need to take stock of your emotions; however, you need not always enter into them. With the help of perception, you need to put yourself in your partner's shoes to see things from the other side of the table.

Most type four's tend to be highly sensitive and emotionally complex. However, you will need to be specific and direct when it comes to things that you want and does not want. If you feel offended, it is understandable that you are angry; however, you need to be careful not to blow what others say out of proportion. Instead, try to understand what the person really meant.

You should have a strong friend circle. You should never have only a single friend who is the support system of all your emotional needs. Additionally, you should appreciate the gifts that your friends are bringing to the table. Make a note of saying out simple gestures like thank you, etc. These small tokens of appreciation will go a long way in making friends who are much more meaningful than those who are there temporarily.

You need to strive to deal with all your interpersonal problems immediately rather than stalling or withdrawing from them. You need to understand the world runs on perspectives; hence, you need to be careful so that you do not overpower the people with your words and emotions. At times, you might feel the overpowering surge that you are correct; however, you need to take control of that urge and refrain from speaking. Listening is a great motivator.

If you are feeling like you are about to have a really bad mood shift, make sure that you are away on your own for some time. In these swings, there are chances that you might blabber out something your partner or friends may not like.

Emotions and feelings

Type four possesses the unique ability to see things from a different point of view and offer a higher perspective on matters while the foot remains firm on the ground. However, when extreme feelings and emotions are involved, the judgment might get clouded. For dealing with emotions and feelings, you can write them all down in the harshest language possible in a letter that you will only keep to yourself. Whenever you have trouble controlling your emotional side, you can do the following:

You first need to perform a loving act towards yourself each day. Get detached from everything and become a silent observer. Visualize the situation where you are slowly turning down the heat. Then keep noticing what is being missed and learn the value of positive thoughts of what is here at the present. Stand still and observe everything; remember that the situation and feeling will pass.

Whenever you feel like you are about to burst out your emotions, you need to take a deep breath and maintain your focus. The classic sign of 'envy' is more prevalent as an emotion in type four's than in any other enneagram type. The feeling that other people have something that you don't bring up an aggressive counter-reaction or deficiency. However, envy is only a part of something much bigger.

The feeling of being disconnected from the Holy Origin, or the source, is a part of human nature. However, type fours tend to experience this feeling more deeply than others. This is because it is a part of their nature and leads them to a quest for something. But once this quest is completed, the deep connections and the wholeness turns into melancholy and loss. Because of this, you tend to feel vulnerable. Always ensure that you have something or the other to do to prevent your mind from flowing in different directions.

You have to avoid postponing things until you are in the 'right mood'. If you want to, commit yourself to the action and you will be able to do meaningful and productive work that will contribute to your own good and the good for others.

Additionally, you should also stay away from having long talks with your imagination, especially if they are not giving you good thoughts, excessively romantic, or resentful. Instead of living under imaginary relationships, you need to actually live in them. With the help of this method, you will not be entirely operated based on your feelings. Mostly, type four's has the tendency to amplify their feelings way out of proportion. In this case, you need to check if the feeling is coming from its intensity or is it genuinely that feeling.

While you might have a big capacity in your heart, the non-acceptance part of like will always create a contraction. You will have a lot of feelings, but still lack the kindness towards yourself habit is strong. For you, you need to combine the quest for wholeness and the longing of your heart together to form self-acceptance. Most type four's feel that the evoking feeling of gratitude balances this feeling that something is missing.

Avoiding depression

As you are progressing on a spiritual journey, you will start to realize the significance of emotional steadiness and the peace that it brings. Your search for stability paves the path for your capability to sit with sensitive conditions without dramatized narratives. You need to let go of the past and come to the fact that it is over. Via forgiveness of self and others will you able to tap on all the anger that is hidden underneath you and work towards fixing the sadness.

One of the most powerful and basic desires of this type of enneagram is the longing to find something significant in almost everything. This is the reason why fours are considered almost always compassionate and creative people. This particular trait helps them find something beautiful in almost all painful circumstances. However, this particular desire also leaves them depressed, lonely and hopeless because significance cannot be found in all things. In this case, you need to get out of your house whenever possible. In fact, you should visit far-off places every once in a while. This will offer you a refreshing feeling and a new perspective in life.

For fours, their feelings are everything. They motivate you and help you make decisions. They also feed your creativity. While this connection of emotions and feelings is a beautiful trait that lets you access feelings like compassion and empathy, it can possibly also ail the filter through which you experience life. Some triggers of depression are based on what you feel about something, rather than what is going on right now. Since you are already fully immersed in your feelings, you will have a tough time getting out of your space and embracing the standard perspective. While your feelings are quite real, valid and might bring insight, they are not very reliable.

In this case, it is always helpful to learn more about yourself and the qualities that might lead to more difficulty. You need to move towards goal-oriented objectives and do things like yoga and exercise. While these goals are quite small, they are also quite meaningful and will help you in coping up with depression.

Because this group of people is not able to live up to their idealized self, it gets quite difficult for them to appreciate all the positive qualities they actually have. In reality, you are secretly desiring for someone to come in and rescue you by bringing out the core person that you are. However, you need to learn to let this idealized version of yourself go and appreciate the unique individual that you are.

Although comparison, anger, and jealousy are traits that affect all enneagrams, type fours have unique internal motivations and patterns that make anger and jealousy a more initial scuffle. This mostly happens due to two reasons – fours have a desire to find something unique in everything and the fear of missing out on something. This combination contributes to a lot of anger. If you are upset about a person, you need to confront them so that you do not turn your anger towards yourself. It is okay if you want to mourn major losses. If necessary, get professional help as well.

You need to develop a lot of good habits like working, eating, exercising, and sleeping on time. While work is important, you also need to stay in touch with your friends, families, and relatives. These things might seem like simple solutions, but you will be surprised by how much positive effect these will have on your life. You need to talk to your friends and family if you feel like you are heading under depression. Depression is a disease that can be cured if treated right. Alternatively, you can express your depression in a creative manner;

there are several forms that can help you do exactly that like art, dance, music, poetry, etc.

You should always look towards the positive aspects of your life. Positivity will help you look at the brighter sides of life. If needed, you can write down everything that you are thankful for and post it on your wall. This will remind you each day about the positives that you have inside you and that it is enough to overpower the negatives in your head. You should know that negativity is inside everyone, and it is possible to suppress these thoughts and feelings.

Things that Fours' would never do

Paint a landscape and hang it over their sofa in the living room. They would also never visit a restaurant in a polyester suit and order an instant coffee and a Spam sandwich. They would think about nothing but the past achievements and positives that have happened in their lives. Additionally, deciding that indigo, champagne and mauve are not really great colours and coming up with alternatives like Laughing Yellow and Happy Pink. They also do not give up on soul-searching because it is too tiring or throw away old photos of lovers in a dispassionate manner.

Positive things that you can tell yourself

- I will be kind to everyone in the same way that I am to my best friend.
- I am lovable, capable, and beautiful; I love myself for what I am and there is nothing that can change that.
- My body is my home, world, and universe. I promise from the bottom of my heart that all is well.
- The present time is for relaxing and enjoying.
- Each day I will value, no matter how imperfect it might get.

BONUS CHAPTER

How To Get Along With Different Enneagram Types?

How to get along type one?

Enneagram type one, or also known as a perfectionist, is always on a lookout to make things better since nothing is ever good enough for them. This particular pattern of thinking makes them perfectionists who have the desire to improve or reform; they tend to think of themselves as idealists who want to make order out of the omnipresent chaos.

Type ones have a fine eye for details. They are self-aware of the flaws they have in themselves, the situations and others. This awareness triggers their need for improvement, which can be beneficial for the people involved. However, it can also prove to be quite burdensome for the type one and those who are on the receiving end.

Type one lives to make the world a better place to live in. Ultimately, this obsession also turns out to be their greatest fear; this fear is that they are innately bad, thereby pushing them to always improve themselves. When it comes to a relationship with family, friends, and lovers, type one's, when on the healthy side, are easy-going, care-free, and fun individuals.

However, when they are not in good health, they tend to become overly resentful, stubborn, and critical towards others. The relentless pursuit of the ideal can make type ones have a hard time relaxing. They will often deny themselves the simpler pleasures of life.

If the concerned partner is a type one, you need to show your love for them by taking an interest in the things that they are passionate about. That type one person is only trying to make things better for you. They have a strong intuition and sense that can improve the lives of the individual and the whole world. If you can achieve this, you will be appreciated by your type one friend/family member/lover for your part. Type one will listen and value your advice and you will be able to bring a smile to his/her face when you acknowledge their achievements.

When you are relating to type ones, you must remind them that they are innately good people. Since it is in their habit to see flaws everywhere, you must shift their way of thinking and learn to give them a lot of kindness since they are trying to stop their inner critics.

Type ones also try to stay in the cleanest environment possible; so, it is appreciated by them if you honour and respect their need for cleanliness and order. More than anything else in the world, you need to be very gentle with type ones, especially when they are stating criticism. You should never take anything they say on a personal one. Even during the roughest day, you need to remember that no matter how many types one is critical towards you, they are usually much harder on themselves. You need to take a share of their responsibility so that they do not end up with all the work. You also need to be considerate and fair to them, apart from stopping them to be very hard on themselves. You need to encourage them to lighten up and laugh alongside people when things tend to get uptight.

Ultimately, type ones will start appreciating your gestures when you gently encourage them to lighten up and laugh about it when things do not go exactly how they planned. You need to remember that type one are the hardest on themselves. You need to remember that when they are correcting you, it is simply one of their ways of showing their love and affection for you. You need to remember to encourage them to be gentle on themselves and with others.

How to get along type two?

The enneagram type two, or also known as the helper, tends to feel worthy to an extent as they are helping other people out. For type twos, love is the greatest ideal. They strive to give to others selflessly. They are emotional and warm people who care a lot about their relationships and will happily invest their time and energy to the people they love and will appreciate their efforts. Type twos also know how to make their homes inviting and comfortable.

Type two's are naturally intuitive to the requirements and needs of others and are considered the most caring and empathetic out of all enneagram types. For type two's, relationships are everything. They do everything to build connections and identities are established based on the levels of interactions. When at the peak of their health, they are capable of balancing self-care and care for others.

On the harder side, type two's tend to develop a sense of entitlement, especially to people who are closest to them. Because of this, they start to think that a certain amount of gratitude is owed to them. They start becoming demanding, intrusive, manipulative, and bossy. When type two's feel that they are not receiving enough love and respect for their hard work and efforts, they can become irrational, hysterical, and even abusive.

If you know someone who is a helper, you need to show your love for them by letting them know how much respect and love you feel for them as real and honest as possible. You should always try to give them specific examples of all the things that you appreciate about that person. Since type two's tend to live directly through you, you need to ensure that you are helping the person pay attention to their own needs and acknowledge their problems with compassion.

You need to encourage the person to take care of themselves. You must let the type two know that it is perfectly fine and also important to take care of their passions and interests as well. However, you need to exercise caution while criticizing type two's because they can be very emotionally sensitive. Be appreciative and share some fun time with them. Let them know that they are special and important to them. Even if you decide to criticize, do so in a gentle manner.

You need to teach type two's how to love themselves. You need to constantly remind them that life is not an endless well and their identity need not be wrapped around who they are and whom do they care for. But it is not only about asking a type two to stop and take care of themselves; you also need to help them pave a path towards self-care.

Since type twos are mostly verbal processors, you need to ensure that you listen to them perfectly and engage them in conversations for a hale and hearty relationship. Because they fear to let other people know about their wants and needs, they need people to fight for them and remind them constantly that they are not going to leave them anywhere.

How to get along type three?

Also known as the achiever, type threes are the type of people that require validation for them to feel worthy; this comes in the form of success and the need to be admired. This group of people is highly focused on their goals, competitive and hard-working; the goals may vary from being the best salesman in their company to the most wanted person in the social group.

The people in this category are goal-setters and know how to achieve them. They also know what drives a person to reach for the stars. They are highly competitive and always on the lookout for recognition and praise. When type threes are deeply engaged in their work, it is always best to leave them alone.

Type threes are usually charismatic, often extroverted, and socially competent. They have a good idea of presentation and often give out an aura of the zest of life that can prove to be contagious. They have good networking skills and know-how to reach the highest ranks. However, under all these qualities, type three has one downfall – they are afraid of being called 'losers'.

Achievers are also known as shape-shifters and are willing to do any type of role in society, as long as they are celebrated for their success. The negative aspect of this is that they tend to lose touch with the real selves and start living their fantasy land with reconstructed images as their authentic representation.

Because of their need to win, relationships with type threes can feel dishonest and fake, especially when they disconnect from the present moment and their feelings. Additionally, they also find intimacy difficult at times. Their need for external validation is often hiding the deep sense of shame of who that person really is; type three shames that this fear will be revealed to the person if he/she gets too close. On their worst, three's sense of narcissism can get very ugly and they can become ruthless and cold-blooded in the pursuit of their goals. Let them know that you like being around them and praise them for their achievements every once in a while. When giving feedback, make sure that it is honest, not judgmental or unduly critical.

If you know someone who fits into this category, you need to let them know that you are proud of their achievements. You need to be very specific as well. You should give off an aura that conveys the message that you love to be around that person. Give away subtle compliments regarding their looks and how others see them. Slowly, type three will start warming up to you and will appreciate everything that you have done for the person.

Additionally, type threes have an urge to create a façade that they think you might love. They simply do this because their idea of self-value is not too high and think that their performance will make up for it. Hence, you must discourage masking and shape-shifting whenever you are with a type three. You need to acknowledge their successes, but not justify their worth with these achievements. You need to vocalize your love and respect for them and discourage them from creating these constructed personalities.

Type three is most appreciable when you allow them to work alone. However, you need to remind them from time-to-time to slow down as there is always a risk of them becoming a workaholic. You need to keep their environment peaceful and calm. As time moves on, type threes will appreciate the honest feedback that you offer; however, ensure that you do it with compassion and care since they can extremely defensive when provoked. You need to be gentle.

Additionally, you need to remember that type threes are very focused on their future and can sometimes seem distracted during conversations. You should not take these traits on a personal note. Also, they dislike living in the past or dive deep into emotional feelings. Remember to give them space while also encouraging them to practice healthy boundaries; also, engage them in non-competitive activities.

How to get along type four?

Type fours are known as romantics. They tend to experience a sense of melancholy and longing. Something is always missing for them, which then leads them to a quest for wholeness via romantic aesthetics, healing, or idealism. They have a firm belief that lost love can be regained by finding a situation or love that is fulfilling, special, and unique.

Type fours live for relationships and connection, even if they believe that they are not. Most four's tend to believe that they are too messy, complicated, or too much for most of the members of the society; however, they are always relentless when in the pursuit of an authentic connection. Type fours seek depth and meaning in their quest for personal creativity, work, and relationships.

You might notice that most of the artists fall under this category; they are known to give away moving and profound creations to their communities. They excel in expressing basic human emotions in the form of poetry, music, and dance. While a good image is important to them, the higher priority still remains authenticity.

Type fours are highly sensitive and complex. They long to be appreciated and understood. While at their best, type fours can help other people learn how to engage with their feelings and lean away from the dark corners. The challenges only come when people trust their emotions as the truth, primarily because their feelings will often tell them that they are flawed. As a result, type fours can become unappealing and moody, thereby detaching themselves from the world and spiral into melancholy.

Type fours are always tempted to conform for the sake of acceptance and belonging. They tend to live in worry; while they want to be accepted into society, they also wish to stand out from others. These worries make them think that there is something wrong with them and they get easily convinced that relationships and genuine connections are not their cups of tea. They get envious of other relationships.

You need to remind them that ordinary moments are just as attractive as the astonishing ones. You should comfort them and let them know that it is perfectly normal to be misunderstood. It is also absolutely normal to have a small circle of intimate friends rather than have a large circle of fake friends. If they think no one gets them, you need to convince them that it is perfectly fine and every person is unique from the other. Give them a lot of compliments, especially because it will mean a lot to them. You need to be a supportive partner or friend and help them love and learn from themselves.

Forging a relationship with type fours can be quite an overwhelming experience, especially if you are not ready for the emotional complexity. However, if you can stick with it, it will be worth everything. Out of all the enneagram types, the romantics are known to offer gift a space for engaging with feelings and profound connections. Stop telling them that they are overreacting or are too sensitive.

Type fours know how to hold on to despair and joy at the same time, which means that they can seem extremely moody and overly sensitive in one go. However, you need to take care so that you do not point it out to them. They know that they are capable of experiencing all emotions and they prefer to keep it that way. You need not match your emotional connect with them; all you need to do is be true to who you are. There is nothing more that means anything more to type fours other than authenticity. Try not to minimize their feelings or conceal yours. They prefer single conversations rather than group gatherings. Always respect them for their gifts of vision and intuition.

How to get along type five?

Type five, or also known as the observer, are the type of enneagram that focus more on accumulating knowledge and intellectual understanding. They are often known as technical experts or scholars because of their analytical abilities and perception. They also tend to be self-reliant and private. For them, relationships might prove to be a risky affair and it can be quite difficult for them to open their hearts out and share their feelings.

Type fives, on their best days, are known to bring in a lot of strengths to any type of relationships like intellect, curiosity, insight, and vision. They tend to feel a familiar comfort at home, where they are amid their thought realm. Generally thoughtful, well-read, and intelligent, they quickly become experts in the areas that align with their interests.

Fives are often showcased as eccentric; they feel like altering their beliefs to accommodate the opinions of the majority. However, they will not compromise on their freedom to think whatever they want. The main problem with type fives is that while they are really good at thinking practically and make calculative decisions, they are equally bad at dealing with the demands of their emotions or a relationship. Because of this, they tend to be reluctant when asking for help, independent, non-intrusive, and shy.

If you know someone who fits into this category perfectly, you need to remember to show your love and respect about talking about something that they know best. You need to find out what they are most passionate about. Once you do, you will see that the conversation will last for hours to come and both will love each moment of it.

Since this category of the enneagram is quite shy, you can take the type five-person out for a nice dinner instead of inviting them to a party where he/she will not know anyone. If you do, the person will love and respect you for it. You need to be independent, not clingy.

Make the person feel welcomed, but moderately, or else he/she will think and doubt your sincerity. If they ask for it, help them avoid intrusion of their privacy, overdone emotions, loud music, and big parties.

You need to give them a lot of space. Mostly, they are introverts and need some quality alone time to recharge their batteries. If you wish to converse with them regarding serious matters, you should always give them the time and space needed to respond. The person will appreciate your patience. Additionally, you should remember that the person is not a robot; so state information in the most concise and simplest manner.

Type fives need a gentle push to engage in their emotions. They have a tendency to revert to their old thinking style. So you need to keep pushing and nudging them; also, you need to constantly remind them that their vulnerability should not be seen as a burden, rather a gift.

You need to remember that type fives tend to keep a small friend circle. This is because they have limited energy and they tend to spend time on the ones they are closest to. They can be quite sarcastic and are comfortable with their life rhythms and stimulating routines. You need to foster a long-lasting relationship by nurturing and respecting these traits.

How to get along type six?

Enneagram type six, or also known as the questioner, are security-oriented and committed type. They are trustworthy, responsible, hard-working, and reliable. They have the capabilities to predict problems and foster corporation. However, they also tend to become anxious, evasive, and defensive; they might also get the job done on stress and keep complaining about it. While at their best, they can be courageous, self-reliant, and internally stable while championing among themselves and others, they tend to become suspicious and question themselves.

Type six make use of their intellect and perception to understand how the world works and find out who all are hostile or friendly. They primarily divert all their focus on guarding the safety of their community, project, or group. They have the ability to anticipate a particular problem and come up with solutions accordingly.

Type six people are only looking to feel secure, to be supported by their friends and family, to have reassurance and certitude, and test the attitudes of other people around them. Often, you will come across them as hard-workers and energetic. They only wish to listen to the truth. They do not act lightly to people who are trying to flatter or manipulate them. They have a desire to know where they stand and where other people stand so that they can act accordingly.

Type six is also known as anti-authoritarians and will suspect any person who will display power over them. For this reason, along, they might feel reluctant to assume the role of the leader and lead the team themselves; they simply want to ensure that other people will not resent them when the time comes.

If you wish to get along type six, you need to be as clear and direct as you can. You should be willing to give in a listening ear carefully and hear everything that they have to say. They also tend to get a lot anxious; so, you should never judge them for that. You simply need to reassure them and let them know that things will eventually work through.

Reassurance is really how you get on the good side of any type six. Some ways to do this is to make jokes and laugh about them with the six. You should never push them into learning new experiences; instead, coax them gently and let them know why they should opt for a certain activity. You should never try to react, especially when they are overreacting. You need to clear and direct as well as listen to them carefully. Never judge them for their anxiety and keep reassuring them that everything will be okay.

You also need to be trustworthy and consistent to get close to type six. You should be capable of disclosing your own personal thoughts and feelings to them. You should appreciate their attention to your problems and agree on the rules and procedures. You need to join them in accepting what could possibly go wrong and help them move ahead. You need to stop being ambiguous and put as many cards as you can on the table. Do not overreact on their overreaction.

How to get along type seven?

Type sevens, or the adventurer, make for wonderful co-workers, partners, and friends. They are always optimistic and fun, light-hearted spirit and always on the lookout for adventure. They know how to connect with their inner child and remind others how to play.

However, while sevens might appear to be having a lot of sunshine and fun, they often struggle when it comes to confronting their feelings, especially when they think that their feelings will be portrayed as negative. If the sevens are at their best, they can be seen as a beacon of hope, like Superman. They wish to see a safe world and the best in people. However, when hard times fall, they can become very opinionated and hard-headed. They will start failing to see the finer details and will rarely follow-up on their work and commitments.

If you know someone who fits in this enneagram type, you will be able to experience all the joys of the relationship. However, you need to remember that type sevens are always on a pursuit of pleasure that can make your presence less viable. They are very much oriented towards the future and will always want to seek the best experiences in life. When pain and hardship comes, they will try to sit with those feelings. You need to help them regain control of their minds and reframe the negative experience as a learning one.

Similar to how type sevens are capable of teaching other numbers on how to enjoy their lives, you can teach them about holding their emotions. Most sevens have the tendency to think that they have a simple lifestyle with little or no feelings at all; you need to convince them that the statement is false. You need to convince them that all human beings are complicated and fitted with a wide range of feelings, which is the best gift that is given to us all. Give them freedom, affection, and companionship at all times.

Since type sevens are quite natural when it comes to hiding negative emotions, it is quite natural that their pain can manifest into shame or anger. Additionally, as they are driven and energetic individuals, they tend to keep aloof most of the time in order to balance out the time that they spend with people. They do not do too well with co-dependence and can feel trapped either by the expectations and need for others. You need to engage them in their favourite activities and hobbies for a healthy relationship. Do not try to change their style.

Type sevens will love you to join them in their adventures of life. The person will definitely enjoy your company if you tag along. You should not dwell in life's difficulties and type sevens will make sure that you pay attention to the good things in life. However, you should not drag the type seven down since the concerned person does not like dealing with the unpleasant aspects of life as well.

How to get along type eight?

As the name suggests, type eights, or the assertors, are leaders who believe in taking action. You might notice some traits like that they always want to take charge and are always on a lookout for solutions. They are highly-energetic and wish to find meaning while standing up for the underdogs. When at their best, type eights are generous, playful and supportive.

However, if things are not going the way as expected, they can become quite combative and aggressive. They find it extremely difficult to relate to feeling-driven numbers and thoughts. At times, they can be misunderstood as bossy characters or bullies, especially if the type eights are women. Since they fear that their emotions will control them, they tend to distrust them, whatsoever. They will also avoid displaying any type of vulnerability that will expose their weakness.

If you are having a hard time dealing with someone who is a type eight, you first need to come to their level. They value straightforward communication and honesty.

You need to remember that their aggression is not personal at all – instead, they are trying to protect themselves and seeking to control the environment. You need to stand up for them and be direct, strong, and confident. You should never gossip about them or betray their trust. Give them ample space when they ask for it.

You need to stay true to whom you are if you wish to have a healthy relationship with type eights. You cannot force them to show their vulnerable side because it is in their nature; however, you can create a closed and safe environment for expression. If you respect them, you will find yourself in an incredibly stimulating and deep relationship. Even if you think getting this close to an individual is not okay, you do not have to worry about anything. Most type eights only make room for a small circle of intimate relationships. If you do not find a place in this small cycle, it does not mean that they do not like you. It is just that they might have maxed out on the number of people that they can have an emotional connection with.

You need to show your love by being upfront and honest. You should never intimidate a type eight. At times, you might notice that the person's attitude might be mistaken as aggressive and pushy. However, you need to remember that these are the very same traits that make the person sensitive and vulnerable. You need to help type eights let go of this emotional wall, after which you will notice their tender side; they will be grateful to you for your unconditional love.

Type eights will also appreciate your efforts when you stand up for them in their time of need. All you need to do is appreciate type eights; remember not to overdo it though, since they dislike flattering. While their loud voice might come across as intimidating, it is usually how they are.

How to get along type nine?

Type nines, or the peacemakers, are also known as the chameleons of all the enneagrams. They expertise in relating and adapting to all the other numbers. This can be seen as both, a strength and a weakness. While this enneagram type knows how to make everyone feel present and seen, they also tend to lose who they are in the relationship.

Type nines have a tendency to cling to a lie that neither their opinions nor presence matter. While they seem like easy-going characters, they tend to erase themselves to maintain peace. Some main challenges regarding type nines in a relationship are that they tend to become passive-aggressive, aloof, and distracted towards others. However, if you manage to catch them on a good day, they know how to avoid conflict for connection and will ensure that their opinions are asserted when it matters the most. The most unique aspect of type nine is that they tend to forget their autonomy while validating other enneagram types.

When you are relating yourself with type nines, you need to remember some major pointers. To start with, you need to always include them in the decision-making process. You need to encourage them to voice their opinions since they will naturally move away from the conversation. You need to give them choices when asking questions rather than open-ended ones. Type nines also look for affirmation when they are being honest, so you need to make sure to celebrate when they do speak up.

Another aspect you need to remember is that type nines are mostly passive-aggressive in nature. Hence, they do not explode their rage like other enneagrams like eights. Instead, they will let that rage get buried in themselves or let it out in an indirect or non-verbal way. You should always aid in fostering a positive expression by giving them time to process and cultivate a non-violent environment. If you wish for them to do something, you need to ask them in an appropriate manner; they dislike pressure or expectations. While you can be of service to them, you should never try to take advantage of them. Listen to them closely and let them finish talking.

You need to remember that they might not be as direct in conversation. With time, they will get comfortable and will share their thoughts and feelings with you in a safe environment. However, if you feel like they are simply burying their emotions, you need to remind them that this habit of theirs will eventually take a toll on their health. Healing and making profound connections with other people are just as important.

CONCLUSION

Working with the enneagram to lead a more fulfilling life

Although we first heard about the enneagram teachings in 1915, it is still popular today due to the benefits it can have in various aspects of our modern lives. It not only allows us to develop a more in depth understanding of self-awareness but in addition, it aids the understanding of others around us. Consequently, studying the enneagram helps us to free ourselves from our behavioural patterns, which distance us from our essence. Gurdieff, In fact, talked about the differences between the true essence of a person and their personality and how the true essence of a person is rarely manifested because our personality, which is developed from external inputs, prevails. The enneagram teachings are studied to aid us in escaping the loop of behavioural patterns that we have learnt.

The enneagram does not only benefit individual people, but also helps teams work well together and aids organizations. By increasing an individuals' empathy towards others and allowing them to understand the motives behind some peoples actions, it consequently improves teamwork environments as people who have studied the enneagram will tend to criticize others less.

The enneagram helps us obtain the best version of ourselves based on our personality traits. Each personality type has a different way to perceive the World around them and what triggers how we think, feel and act. In our daily lives, we have automatic reactions to different events and the Enneagram system enables us to pinpoint some of our limiting behavioural patterns. Despite being predictable, a person's automatic reaction to a specific event is usually difficult to modify because they are unaware of this limiting pattern. By identifying our personality type, we are bringing awareness to our triggers and unconscious behavioural patterns, and can begin our changing process and manifest our true essence instead of our external personality. Our Essence does not need to be "found" as we already have it; it is just hiding underneath our personality or as it is sometimes referred to as our ego. The Enneagram could help us on our path of unfolding our essence, rediscovering what lies within us, remembering. This process can also be described as spiritual awakening

Our internal character and how we perceive the World, is based on this main centre therefore, the key to bettering ourselves is held in our centre. Having the understanding of which centre we mainly operate from, is vital when taking the steps towards increasing our self-awareness. Once we have the knowledge of which centre is stronger within us we need to integrate all the three centres within ourselves to enable a more balanced life, and consequently improve interpersonal relationships.

Our main centre that we use to interact with others is often unbalanced in relation to the other centres. A person who is heart centred, for example, will tend to act and make decisions based on an emotional input without taking into consideration gut feelings or logical thinking behind the situation. Similarly, somebody who is mainly body centred will act on impulse to a situation without noticing how their behaviour or decisions will affect others, and they will not stop to think logically, in most cases. A head centred person, on the other hand, is likely to over analyse a situation, going over and over the details in their mind, making it difficult for them to act. Therefore, when you know which your strongest centre is, it is useful to observe your reactions and interactions and try to work on listening to your other centres so as to balance them out.

The Enneagram is used and worked on to help our journey of self-understanding. One of the way we can use the Enneagram process to work on our inner self is to bring our attention inwards, to transition our outward attention and bring it inside. A three-way centre mediation can be useful for this transition, simply paying attention to breathing and feeling and accepting any sensations you observe, trying to concentrate on the three centres separately and being aware of where your attention gets stuck, eventually trying to get it unstuck.

After having taken a more detailed look into our personality traits and characteristics and begin to open up and realise our flaws and our problem areas or behavioural loops, we can start to question the negative side to our personality and analyse how we interact with others and why. We can also use the enneagram personality types to learn how to improve our relationships with other types or try to understand why others act the way they do towards us.

CPSIA information can be obtained
at www.ICGtesting.com
Printed in the USA
BVHW051148050821
613733BV00003B/452